FARMER PIRATES
AND
DANCING COWS

Farmer Pirates & Dancing Cows
Lynn R. Miller
Copyright ©2006 Lynn R. Miller

Publisher
    Davila Art & Books with
    Small Farmer's Journal Inc.
    PO Box 1627
    Sisters, Oregon 97759
    (541) 549-2064
    www.DavilaBooks.com
    www.SmallFarmersJournal.com
Authored by Lynn R. Miller

First Printing First Edition 2006
ISBN 978-1-885210-31-9

Also by Lynn R. Miller
    Art of Working Horses
    The Harness Book
    Haying with Horses
    The Horsedrawn Mower Book
    Horsedrawn Plows and plowing
    Horsedrawn Tillage Tools
    Training Workhorses / Training Teamsters
    Work Horse Handbook

    Ten Acres Enough / A Farm For Free
    Old Man Farming
    Why Farm
    Talking Man
    Thought Small: Poems, Prayers, Drawings & Postings
    The Glass Horse
    Brown Dwarf
    Roots in a Lovely Filth

Front cover photo: Lynn Miller returning his horsedrawn mower to the field.
Photo by Jean Christophe Grossetete.
Back cover: Lynn Miller and daughter Scout Gabrielle enjoying a garden harvest
of new potatoes. Photo by Kristi Gilman-Miller.

dedication

for my children
Maxwell Henry, Ian Lewis,
Justin Everett, Juliet Christene
and Scout Gabrielle

you have always
kept my eyes
on the prize

table of contents

Acknowledgments

*Special thanks to Susan Tank, Amy Jo Ferris,
Judith Hoffman, Marcia Plows and Kathy Blann
for their many hours of typing and proofing.*

preface

This is a book of essays. These essays came in and on their own good time and terms, scattered over many years. It was long felt that these and many more like them belonged together in a volume (or volumes), as they all approach the same central view of a life of farming and a farming of life. And they come of this one man's peculiar context; a context which puts farm work and inquiry at issue with modern (and post-modern) times.

For decades closeted thinkers and store-bought scientists have held that the world, our world, had to be de-spiritualized, de-animated; denied the capacity of subject. This is what modernity has stood for, a so-called declaration of reason's independence from mystery and magic. The evidence of what harm this has caused the family of man and humanity's nest is everywhere around us. And, for me at least, nowhere more poignantly oppressive and depressing than within industrial agriculture. Knowing this and fearing the consequence, I have chosen to work towards re-enchantment.

Why wouldn't I, in moments like this, lean on inquiry, lean on thinking through to some new shelf? The working, as in days like yesterday - packing bags of powdery-white, burning lime to the spreader and doing pre-winter ranch preparations - always cleans my pipes, just as these inquiries would sometimes gum me up. Between the clarity of the work and the eventual depressions from thinking lies an ocean of vitality all mine. When the working is strong the nights seem to pinch off separate from the days. Dreams pile up behind a blanket of thick postponement. All of the morning's first moments go to the shake off of *slow,* as energy winds towards the anticipated efforts. I long for the fencework and the field cleanup. My dessert is the painting. My purpose is in the meld of work, creation and clear thought.

Coming to this life of mixed Wisconsin/Puerto Rican

heritage, raised in suburban and urban centers and powder-coated by the occasionally useful immersions of public education, I was confused. There was for me the slow melt of the nineteen-fifties into the sixties. Perhaps that is why I felt ambivalent to those flame-thrower urgencies which fueled the onslaught of the atomic and computer ages. In there somewhere is the reason why passions which pushed me have always been centered on the widest canvas of the archaic arts: music, dance, theater, painting, sculpture, writing and ... farming.

For the French poet Mallarme, art for art's sake was religion. For me painting has always been for painting's sake, and farming for farming's sake. Writing, however, has been quite another thing. I have never been able to get a hold of the idea of writing for writing's sake. I write rather as a form of thought transport, as a conveyance, I write, hastily, to get somewhere with an idea or an argument. For this reason I have been want to take the craft of writing serious. If livestock genetics and cadmium pigment properties are known and valued by me, it has been by passionate interest and choice. With writing I have no similar earned respect. Long ago I gave up on the idea that I had what it took to be a good writer. But I know I have what it takes to be a fair storyteller. I write, once again, to get someplace, to get a thought out there. I have been doing it this way for forty-five years. Any small successes I have known, as a public speaker and/ or essayist, have come because I enjoy, in those moments, the comic-cloth of poor liar and idiot thief. Hence, one of several reasons for the title to this collection of essays, *Farmer Pirates and Dancing Cows*. Who better to hide behind the dimestore costume of marauding agrarian than a child of both Wisconsin and the Carribean? Who better to champion the silly notion of bovines in rumba but a farcical farmer?

And piracy? To my view *the gold in thar ships* is "relevance." We commoners are not to have it, let alone understand it. It belongs to the marketplace, to the patent holders, it's the secret behind their control, power and wealth. As farmers, when we deign to ignore the "rules" the marketplace puts upon us we steal their relevance. We use their seeds without permission and royalty fee, we squeeze eggs out of our unregistered chickens without an ounce of quota allowance, we milk our sweet cows without so much as a howdy doo. For these crimes and more, the marketplace would label us as pirates. So be it.

But dancing cows? Goes right back to that notion of farming for farming's sake. Many of us are in the "business" because we love where it takes us, what it would make of us, what it shares with us. From the sublime to tragic, from the fruitful to the barren, all of it direct connection to the humor and humus of the cosmos.

Gathering together all of the nearly 200 essays I have penned, and then paring the pile down to those twenty-two which might feel like a family of thought, has offered up to me the surprise that as a whole these essays do chameleon acts in the fold. The book has become something of itself, something new and different, as if by magic. It is as if reason has lost some of its independence and food has found its flavor.

Please accept this book as one small effort towards re-enchantment.

LRM, Singing Horse Ranch, Fall 2006

Chapter One

The POETRY of SURVIVAL

It is morning. A clear and soft crisp that is unusual for August. The sunlight is insistent and reaches into shadows, outlining the kitten's waiting and lending silver flashes to the streams of milk. The look of the milking does not match the sound of the milking. The sound of the milking and the look of the milking do not answer the fixed determined glare of the kittens.

With my forehead resting on the sweet Jersey's flank, and the automatic contractions of my hands and arms, my mind is set loose -- gradually... first; how so many people have come to view this milking as drudgery? Then, I conjure an image of the hundreds of millions of people around the planet who are this morning milking. And I see the milk in common stream filling one common bucket and I shake my head and chuckle at the absurdity of my thoughts. But something stops me and I see I have slipped once again. Slipped into the trap of apologizing for and even forgiving my best instincts.

The cow's tail catches the back of my head in a gentle sweep and I am back to watching the milk. And then comes the second wave as ideas and pictures come at me from all directions. All of it on the same subject;

SURVIVAL

Not just survival as in staying alive. But as in responsible, stylish, adventuresome, sometimes silly, comfortable, and difficult subsistence.

That's what milking the cow's all about! It is irresponsible, lazy, vulgar, inconvenient, boring and easy to buy a gallon of milk at the store. It contributes very little to your well being. It is, however, responsible, adventurous, and difficult to milk a cow -- plus it can be stylish and comfortable. Otherwise, why is it that folks like to sit and watch me milk? And how is it that I almost fall asleep while doing it?

Survival depends on the desire to survive. If you are to be involved deliberately in the business of survival you must 'want' to survive. If you don't, you may 'remain' for awhile but you won't be surviving.

Sometimes survival connotes "just getting by" or "doing without." The word "subsistence" suggests the same. But it's a mistake to accept these connotations -- they prevent us from seeing possibilities. They are inaccurate mutations of some powerfully important meaning.

One of the kittens seems to come alive as the anticipation fuels nervousness seeking release. A dash across the barn floor and a deathly serious sliding attack on a little hapless tuft of hay results in a comic, rapid-fire staccato of vertical leaps...then! from nowhere, another kitten has thrown away composure to do a life threatening sneak attack on its pirouetting little mate. Oh, the fear and fright and hissing and arched backs -- all of a second -- but gone -- and they both work so seriously to regain their poise. Short-lived as a mama swallow takes a dive and the comedy begins again. This is the pure joy, the reckless youthful abandon of survival school.

Swallows eat lots of bugs. We like having these dive-bombers on our farm. Some people don't like them because of their messy mud-ball nests stuck up in the eaves of houses, sheds and barns. Small price to pay, I say -- But look further, look inside...It's not a "price to pay." And, that is NOT an observation born of the poetry of survival. Swallow nests are a "price to pay" only if you refuse to allow that you and time and all things are in motion -- are passing -- are same.

The life of the Swallow is not threatened by the barn kitten -- it is punctuated by that deadly comic furr-ball. There is a natural equation to that contest which carries no tragedy even in death. But one of us with a long stick knocking down the baby-filled nests because they are unsightly -- that's horrific. We, by such actions remove ourselves from life -- create a deadness within ourselves. Refuse to listen to our best instincts.

It is in this sense I am saddened to see so many people outside of life. This separation is the unconscionable social justification for all the unnatural equations that threaten life on our planet.

Our war systems (I refuse to call it defense), our popular energy systems, our industrial and scientific chemistry, our medical science, our orthodox agriculture, our educational systems, our entertainment, our so called "drug culture," our government -- are ALL outside of life! ALL the result of, or characterized by, an outlook - attitude - posture, that is separate from true instincts for survival. They are all programmed for death in one form or another. They contribute, in their "popular" forms, NOTHING to the quality of life on our planet. And before the horrible threats they pose to our futures, they pollute us NOW in immeasurable ways. All because people amongst us will not allow their own best instincts to govern their lives. That and because of the rule of fear.

We fear the aggressive embrace of a joyful survival because it is to accept mortality. And culturally, as westerners, all systems are geared to the denial of mortality. Here is irony! To enjoy active survival is to acknowledge mortality and in so doing give real opportunity to an immortal character to your spirit and endeavors and produce. However, to hold ourselves separate from the experience of survival in an effort to at least "not have to look upon our own death" this kills us long before we meet our own passing. Kills us by preventing a lifetime of experiencing at least the 'punctuations' of living.

The cow backs up one step and I stop my milking rhythm and hold the bucket. She's finished eating and, with a soft whistling noise, a sighing through the nose, settles into a daydream. I begin again.

Experiencing life right now. That's a big part of the choice to live and work on this small farm of ours. The character of the experiences, good and bad, are forever pulling you awake and alive. Testing.

In a recent conversation I found myself considering the relationship of four words: adventure, luxury, comfort and convenience. "Convenience" is insidious -- it is difficult to know when it has robbed us of value and become an inconvenience, but safe to say often. I confess to being afraid of the meaning of "Luxury." But I have no hesitation saying that for me 'adventure' is 'comfort.' The day to day adventure, for example, of this farm gives me a lasting comfort. And that sense of adventure does come in no small measure from 'knowing' we are surviving.

The milking is done. I get up slow cuz those muscles down low in the back tighten around the vertebrae and say 'old Man.' All of a sudden the floor is alive with six scurrying kittens who all know (or hope they do) what's next. In the corner is a flat pan and I pour some milk in for them. One kitten gets anxious and walks into the middle of the pan getting coated by the milk-fall. It doesn't stop him from lapping up breakfast and the other five set to licking the milk off their brother's fur. No monkey business, no active comedy, just survival -- yet it's hard for me not to feel their little bellies filling with the warm fluid. This is all there is and I am graced in knowing it will always be more than I can hold.

Without experiencing the joy of being alive you cannot survive -- you are outside of life.

Chapter Two

Mrs. Jane's Beans

I need to set the stage: You are a farmer who cares but who also needs to pay the bills. You are a farmer that will go extra miles not to rob or poison your soil, but staying on the farm means coming up with good income. You are a farmer who thinks about things like not selling the fertility of your land but rather recycling it and selling a more finished product. Yet, you are a farmer who is faced, almost constantly, with pressing choices which seem to force "compromise" of values in exchange for the money to pay the bills.

Why is it that this scenario seems inevitable, inescapable, unavoidable? I believe it seems it is because our culture has indoctrinated us to believe certain false premises. These include things like; farming must be treated as an industrial enterprise if it is to succeed, and, the small family farm is a thing of the past, and, agri-business has a vested interest in protecting the future of farming by not allowing rape and poisoning of the land. But the one false premise that I want to discuss, you might not consider in the same league with the aforementioned garbage. That premise is the idea that "marketing," (selling) is an act of combat between bitter opposites. Or, put another way, that you, as a

producer (farmer) must be at odds with your prospective customer (consumer or distributor/buyer) because they want what
you have for less than it's worth. And they in turn feel the same
way, wondering at why you insist on cheating them on quality,
quantity and price. There is an unfortunate old adage born, as the
moral of many western horsetrading parables, which goes something like this: "You've arrived at the right price when both
parties are unhappy, the one because he paid too much and the
other because he got too little." I don't agree with this. I have
learned that it is possible for the opposite to be true and insignificant, all at once. I believe both parties can be well pleased with
the price, and yet feel the price is insignificant, because other
elements in the transaction become more important. And that, in
spite of -- or because of-- the insignificance, true prosperity
becomes the result for both parties. Before you write this off as
some Zen idea of marketing, please allow me some space to
explain.

Markets, and profitable marketing, are so critically important to the ongoing vitality of small farms in our so-called "free
market" system. And yet our attitude towards the process and
participants sort of stacks the deck against us from the outset.
Especially if ours is a caring approach to farming which we wish
to protect. If we resign ourselves to throwing in our cards (and
corn) with mainstream, conventional, marketing channels we
have truly given up. We are at the mercy of some pretty merciless
folks (or, at best, some good folks who are also trapped in the
system and can't give you a fair shake even though they want to).
But the thing that really stacks up against us is our own quiet
acceptance of this notion that selling is a form of bush-warfare
(meaning that we lay in hiding, armed with an unscrupulous
soul, ready to pounce on the first unsuspecting consumer) justified by the assumption that if we don't get these people first, they
will "get" us. And by "get" I mean take our hard-won farm
produce from us for far less than our cost of production. Steal it!
So, given this scenario, lies and deception seem completely
justified ... BUT WAIT A MINUTE! It is a cycle.

Do we honestly believe that everything will be better if we can somehow get the upper hand on people who wish to purchase farm produce from us? That thin-headed notion belongs with the idea that government is going to help us farmers. Those notions belong in the methane-digester!

I want to propose that the prosperous and noble approach would have us consider every potential customer as a fellow human being worth of our trust and respect.

There's a story I need to share with you but names will have to be left out:

MRS. JANE'S BEANS

A young farmer (growing a few acres of vegetables, some oats, hay, pigs, a milk cow and four work horses), desperately needed to avoid the mistakes of the previous season. He had to sell some more of his produce in order to pay back a small operating loan with the local bank. It was winter and he was putting together a planting map and calendar, thinking about varieties and schedules. Somehow he got the idea to dig up some old newspapers from the previous fall to see if the classified ads would disclose some interesting price and supply information to him. Sure enough, the newspaper, for the small city 20 miles away and covering his own small town, had a "market basket" classified section featuring ads for U-pick orchards, locker-beef, and various produce. It was winter, but this farmer was impatient and anxious to at least have a better sense of what he was going to do. He came up with the ingenious idea of placing a classified ad, right then, in that same paper. Here's what the ad said:

CUSTOMERS WANTED. Home canners, big families, order your vegetables now and get big discounts next summer. (Farm name, address, phone).

He didn't get a whole lot of response to the ad but a couple of calls illustrated, rather quickly, that he wasn't well prepared. One woman called and asked what the price would be on a hundred ears of sweet corn. He went to great trouble to answer her questions, explaining that if she were willing to commit to the hundred ears today he would give her a ten percent discount over whatever the market price was come August. She asked if a deposit was required and he said no. She then asked what the price would be if she ordered two hundred or five hundred and what would happen if she bought the one hundred she committed to and then came back for several hundred more. Would she get additional discounts? And what if she got her one hundred and they weren't all good? Could she bring back the bad ones for exchange? Or, what if she got together with several friends and they ordered a thousand ears? How would she look if some of those didn't turn out? He didn't like the way he felt when that conversation came to a close and she said she'd definitely take two hundred ears but they had better be good!

That call, and others, combined to give him a sliding feeling about his idea. He had pretty much checked out of the "committed customer" plan when he got his last call.

(Let's call her Mrs. Jane, she won't mind.) Mrs. Jane asked if he would be growing any string beans and he said yes. She explained that for years an older neighbor man, who loved gardening, had grown wonderful "Blue Lake" beans for her. And Mrs. Jane explained that she canned these beans a particularly lovely way and entered them in several county fairs and in the state fair, winning dozens of blue ribbons. Her neighbor friend had passed away and she was looking for some other source for these beans. But they had to be special. They had to all be picked in the early morning of the canning day. They had to all be a certain exact length so that she could weave each whole bean, in a certain pattern, along the inside wall of the canning jar. (By this point in the phone conversation our young farmer was shaking his head and saying to himself, "Oh, brother!") And they had to be a certain brand of Blue Lake Beans which had a wonderful parallel design on them when picked early in the morning before the sun's heat. (Our young farmer was remembering last year's wholesale green bean price of 17 cents a pound.)

"Well, sir, would you be able to grow such beans? I know it would be a bother but I saw your ad and I just thought ..."

"Ma'am, I could sure grow those beans for you but I confess my specialty is growing healthier food, you know, no poisons and better nutrients. I've never been asked to grow prettier food before. But you know, something like this will take a lot of time and extra care and I'm not sure if I can afford to do it."

(He was thinking, with some chagrin, that THIS was an answer to his discount produce ad?)

"Well, I would certainly be willing to pay for them, how much would you need to have?" answered Mrs. Jane.

"Last year I got 17 cents a pound wholesale. For something like this I would have to charge at least 50 cents a pound." (He thought, "There, that'll scare her off...") "Oh, my," she said. "I'm sure that will be quite fine."

He chokes as he thought to ask, "How many pounds will you be needing?"

"Oh, twenty-five pounds would be fine ..."

Inside, he didn't know whether to laugh or cry. It seemed like the final insult and a really ludicrous comic tragedy. Here he found himself committed to extra work and bother for 25 pounds of beans. Last year he had difficulty selling his two tons of beans!

"Oh, and sir," she continued, "I have one other request, if it's not too much to ask?"

He fought a sarcastic comment and answered, "Oh, that's okay, what is it?"

"I'd like to bring a few gardening friends and come to visit your farm."

He thought, oh well, as least THAT won't have to happen until this summer ...

"Could we come this weekend?" she asked.

By Saturday, he had forgotten about Mrs. Jane. It was 10 in the morning and quite cold. So cold that you could warm your hands by the steaming nostrils of the geldings that were hitched to the manure spreader. He pitched on the last forkful, and went to open the barn gate and saw the light blue Oldsmobile unload its freight of three elderly women. He climbed on the spreader

seat, determined to continue with his work even as he remembered her promised visit. They were at the garage gate as the
geldings pranced out in anticipation of their job. He pulled them
up.

He introduced himself and she commented on how young
he was. He took no offense because he was quite taken with the
comedy of these three well-dressed (one had on a fur coat) old
women wearing rubber boots.

"Oh, my! I had no idea you had these lovely giant horses!
Gosh, they take me back ..." And she gushed and talked to her
friends. He told her he had to go out to the field.

Once out there with the spreader in gear he put the visitors
out of his mind as he thought about the coming year and wondered what it would bring. He didn't know how he was going to
make the payments and keep the farm. All he knew was that he
would try. He was looking forward to the coming year's field
labor like one who thought it would be his last. This in mind, he
turned the team and headed back towards the barn. He was
surprised to see that the three old ladies had followed him out
and were standing in the frosty mud near the edge of the field.

"Young man, can you show me where you will be growing
my beans?" asked Mrs. Jane with a worried look on her face.

Not knowing why, he warmed up to her right at that moment. Something about her absurd, innocent, truthfulness.

"Right there, Ma'am. They will be in the first part of a row
of beans that will go all the way to the fence."

She was wringing her hands. "Oh, my. It's so big. How
many acres are there here?" And she stooped and picked up a
dried and withered corn stalk as she spoke.

"This field, where the vegetables are grown, is 14 acres," he
answered as he watched her push that corn stalk into the soft,
frosty, soil at her feet.

"There," she said, "now I can find this spot when I come
back. Oh, and by the way, my friends would like to speak with
you about some vegetables ..."

It was a wonderful first visit, but only the first. Mrs. Jane bought herself some coveralls and would make a point of coming, unannounced, to check on things. Not out of distrust but because something had clicked for her. She was feeling like these beans she was going to get were a product of a trust she had to earn from the young, naturally skeptical, farmer. She felt a need to contribute at least a genuine concern. And she quickly picked up on the fact that she mustn't get in the way. She brought various people out to visit "her" farm and place what he thought were silly little produce orders. Until one day ...

Spring was coming on fast and he was feeling particularly testy. Small problems made interruptions seem like monumental distractions. He found himself making a most unfortunate speech to Mrs. Jane.

"You know I like you and don't mind you coming around, but I'm having a real problem getting my work done. And these families you bring out on weekends are a real hazard. I don't like worrying about some kids getting kicked by a horse or falling from the haystack in the barn. You just have to understand that this is a business, not some playground or amusement park."

She didn't argue or even respond. She handled it her own way, which was to take the responsibility to protect him from these things. When she came around, after that, she made herself scarce. And if anyone else came with her she made sure they were way away from the work and hazards. And she plotted ways to make it different. Because she became more and more convinced that she should be helping. But she hadn't figured on her health.

Mrs. Jane developed health problems that wouldn't allow her to come by the farm. So as her little short row of beans grew she was only able to come by once to see them. She was over-joyed, not just with the beans, but with the whole market garden, and with her young farmer. He, in turn, missed her visits but the whirlwind of summer farm work kept him from doing anything but keeping a close watch over her beans.

And he did, measuring them in anticipation of the fast approaching morning when they would be picked and delivered. He phoned her and told her that they would be ready any of the next couple of days and she said that would be fine.

But some unexpected problems with the cow's calving prevented him from getting the beans picked for the next two days. When he finally did get them, they were just a little too long for her. Something he hadn't noticed until he had them picked and carefully laid them in the crate, side by side, in perfect symmetry. He took them to her house but she was not at home. So he left a note saying that he would be bringing by another crate of better beans the next day and that she should just dispose of these or give them to some friends to eat.

The next morning he went on hands and knees, with a measuring stick, selecting individual beans from his half acre and carefully laying them in the crate. This time, although they lay side by side and straight, he set a little pattern in the crate that looked like herringbone.

When he got to Mrs. Jane's house he found two surprises. First she came to the door in a wheelchair apologizing for being gone the day before. He wondered at her condition and felt choked, remembering how he had pushed her away during the spring and early summer. And she saw the concern on his face and quickly changed the subject by telling him that yesterday's beans were magnificent and that he had no reason to apologize for them. And, when he smiled and lowered the morning's crate so that she could look in it, he said, "Well, thank you for that. But I think these are closer to what you want."

"Oh, MY!" she said and she started to cry.

She fumbled with her sweater and stuck an envelope in his coat pocket while he put the crate on the sink and asked what he could do for her. That's when he noticed surprise number two. On the kitchen walls were several photographs of him, his farm, and his horses working. And he got the feeling that there were probably more elsewhere in the house. She saw him looking and offered, "Oh, I hope you won't be angry with me. I took them when you weren't looking. I'm afraid they aren't very good but I wanted to have pictures around of "my" farm. Now you'd better get back there because I've sent some people over that want to buy some produce."

He said, "I'll be back tomorrow to visit with you, okay?"

"Only if the farm work is done," she admonished with a soft chuckle.

When he got back to the farm and took off his coat he noticed the envelope from Mrs. Jane. In it he found a wad of money and a note.

"Don't argue with me! It's my farm too!" signed Mrs. Jane.

The amount of money wasn't important except to say that it was far more than their agreed price. And he had forgotten about money from her. It just seemed like she had the beans coming and ought not to have to pay for them.

There is more story that could be told about him, the farm, and Mrs. Jane, but it may be enough for now to say that the year became a financial success for his farm because so many of these customers that Mrs. Jane sent bought and bought and bought produce and insisted on paying what they could afford, which was most often more than he would have charged (and on two occasions less than he would have accepted). And many of them pestered him with their insistence on helping. And he continued surprised at how many of them seemed to take a genuine pride in feeling connected to where some of their food was coming from.

So it was that he found his genuine care for the farm to be an intangible, precious commodity in the eyes of those people who depended on him for some of their food. A healthful, helpful circle.

Mrs. Jane passed away that next winter and left her young farmer fifty quart jars of her winning beans and something else.

That something else was the consummate knowledge that his farm and his farming belonged not just to him but to those others who had become an extended family. From that day on he greeted his "customers" as family with the reminder:

"It's your farm too."

Chapter Three

Tinker and the Fascist Choo Choo

"The concentrations of wealth work against the concentrations of mind. People accustomed to buying things instead of making things often substitute size for the lack of a better idea, or any idea at all."
Lewis Lapham, editor <u>Harpers</u>

The sought-after chance for unseemly and unneces-sary convenience and profit is a gambler's chance carried on the shoulders of reckless abandon and holding more power over the hooked individual than a cocaine addicton. As a wholesale social condition it is the open door through which people forfeit their individuality, their self worth, their security, their sense of purpose. On the other side of the open door is the lizard of corporate fascism secreting the sickly sweet odors of consumption and cultural evaporation.

I am angry.
This is an angry writing.
So what.
An insect in the scheme of things human my voice doesn't amount to a spider's whisper measured against the growl and whine of the whole of mankind's collective greed, need and

consumption. Yet the anger spills out. Must.

This voice of mine is a noise though barely audible. And that is as it should be and how I would have it. For it is the small caring invisible noises of insignificant little people which make up the pure sweet choral sounds of vital small independent communities.

The larger whole of so-called civilized mankind gives off an inescapable stench of decadence and self-imposed significance.

Little communities (rich and poor) thrive, luxuriate and self-pollinate in the effective perfume of accepted insignificance.

Clarification.

We ranch Millers were blessed this summer with a most fantastic long visit from 14 French citizens. Three men and eleven teenagers. They came to see the American west. They camped on our ranch for two weeks. One day, at their request, we took the group to the local county fair and rodeo.

Big crowd, hot Saturday evening, open bleachers, curly fries, corn dogs, aging loud cowboy announcer astride good grey gelding pumping up the crowd to a patriotic fervor but not for 'old glory' or 'these United States'. Nope. Instead his screaming thick "do this with me" plea was for all of us to stand and pay homage to the "Pepsi" riders - six women doing respectable equine drills and each carrying the Pepsi flag. Big flags in red white and blue. I was ashamed.

Boy, it made me mad. I'm not a flag waver myself, most zen-leaning half-carribeans aren't, but I was red-faced mad. At the insult. At the implication! Balled up my fist, shook it in the air and hollered "down with corporate fascism!"

Nobody heard me. Nobody cared. I looked around almost in self-defense definitely in defiance, fist still in the air, and caught sight of one man to my left looking at me with a "yea, right on brother!" look. Only I knew he didn't hear my words. All he picked up on was me shaking my fist. Yet even with un-specified misunderstanding he and I were a momentary community of insignificants.

Lucky for me nobody could hear. Doubtless that had I been heard I would have been canned to death by Diet-Pepsi empties. As for the corporation, I'm just an insect. When anyone is perceived as a genuine threat to a large corporate image, he or she is squashed. My insignificance lends me some protection.

Why am I angry you ask? Surely it isn't because a good and fine "American" company has provided some money to make it possible for the grand old sport of rodeo to survive? (Hear sarcastic well-lubricated gagging.)

You ask how I might connect corporate and fascism. The meanings of words evolve over time and most definitely when parallel but dissimilar circumstances cry for connection as lesson, as warning. The definition of fascism which I chose to apply requires clarification by association and contemporary modulation

fascism: a. any program for setting up a centralized autocratic national regime with severely nationalistic policies, exercising regimentation of industry, commerce, and finance, rigid censorship and forcible supression of opposition. b: any tendency toward or actual exercise of severe autocratic or dictatorial control.

autocrat: one who rules with undisputed sway in any relationship.

collectivism: a. a politico-economic system characterized by collective control esp. over production and distribution of goods and services in contrast to free enterprise b. extreme control of the economic, politcal, and social life of its subjects by an authoritarian state (as under communism or fascism).

It's a new world. I would offer that we are being forced to accept control over our lives from parallel regimes. The first is from our own national governments. The second is from a consortium of concentric autocratic multinational corporations. Large corporations see themselves as outside of national laws and boundaries and actually provide international rule through severe control of the market place. To this end they purchase the compliance of their weaker cousin, national government. In the purest sense, multinational corporations are actually collectives dedicated through market control to rule with undisputed sway.

Witness NAFTA, witness GATT. I use the word _fascism_ because I
see it as such in a modern context and because it gets people's
attention.

No I most definitely don't belong to any extremist organization. Yes I believe in God. No I don't believe the U.S. constitu-
tion is without flaw. I believe in the rule of law. I don't think
justice can be a reasonable expectation in a society ruled by the
rich, the stupid, the suspicious, the arrogant and the lazy. Yes I'm
terribly fond of my country and believe it deserves better than the
odium of prostituted and cowardly politicians media moguls have
seated us with. And most important of all I KNOW things can be
better. I've seen them better. I've been to those pockets of promise and plenty. They are out there now, they are all around us. If
you can't see them, suspect it's because you've succumbed to the
hypnosis of the lizard of corporate fascism. Suspect it's because
you've bought the lie that things are good for you and what you
suspect you miss and want are myths or artifacts. We can break
the spell. We can rejoin with hope, purpose, beauty and fertility.

The entirety of mankind's food supply is at risk.

The health and well-being of the human species is deterio-
rating rapidly.

Weather everywhere on the planet is going cuckoo.

Fertility and biodiversity are being lost to mining.

The planet's destruction is being unwittingly hastened by
misdirected environmental activism.

The exquisite patchwork quilt of diverse cultures is being
replaced by a wash and wear polyester coverlet with a Microsoft
logo.

Smaller societies are being destroyed.

All so called "public" U.S. scientific inquiry is motivated by,
or directed by, short term profit.

The U.S. federal government does not serve the best interests of its people, country and/or planet.

The poor, the hungry, the workers eat each other alive while
an aristocracy of management fill their dumpsters with uneaten

finger sandwiches and chocolate cakes.

And the fourth estate, the news media (print and otherwise), is owned and/or controlled by 6 pea-brained meanspirited billionaires.

The list goes on longer than a senator's dance card, far beyond the Pentagon's payroll and well into the President's defense fund.

And the reason? It is because we have sold our collective soul to big business and signed on to the holy church of corporate fascism. We are all culpable if we are dependent. If we need, for perceived measures of survival, the products they would sell us we are culpable, we are responsible, we are users. If we need the electricity, the petroleum, the processed foods, the drugs, the predigested psuedo-entertainment, the flatulence of government, the sham medical professions, the low-ball schools, we are hogs at the corporate trough.

We enjoy an embarrassent of rationales. I say <u>enjoy</u> in the most vulgar tense.

We, as a civilized culture, rationalize that big corporate agriculture is a necessary evil in order to feed the world. While in truth, "we who have" could give a tinker's dam about feeding anybody, save those who might pay us well for our surplus. While in truth industrialized agriculture destroys our fertility and our biodiversity and poisons our spirit as well as our bodies. And this agriculture we embrace is profit monkey of large corporations we are told we should cherish and obey.

And these "industries" coupled with our obscene rate of consumption are rendering our atmosphere nearly uninhabitable. Anyone who works out of doors, especially farmers, know something is wrong. We have all of us experienced the warmest summer since records have been kept. Mother nature hurls herself at us repeatedly and with growing force. Floods, storms, fires, earthquakes with growing frequency and magnitude and we pretend that each phenomenon is an isolated instance on a natural calendar of weather. We bury our tiny heads in the sand while computerized robots and our odium of politicians assure us that all is better than ever. The U.S. government is paid millions of dollars by petroleum interests to deny we have any atmospheric deterioration. So they do. Petroleum interests do not want us to collectively work on the developmental infrastructure that would make the excellent alternative energy sources, such as animal, solar and wind power, available to all of us. So they buy off our federal government. We do not have a single federally elected official with the intelligence and courage to stand and fight. And the store-bought holier-than-thou hipper-than-you qualified vigilance of most environmentalists is an insult to what human spirit remains because they choose by corporate design and finance to ATTACK individuals, small businesses, farmers, ranchers, fishermen, tree farmers and leave the earth murdering consortium of megalithic multinational corporations alone.

Pepsico and Disney want you to believe it is a small world. It is important to their marketing strategy. It is not a small world. It is a vast, mysterious, magical, unknowable world. To see it as anything less is to buy a ticket for the fascist choo choo. Fast food giants need everybody to want their burgers. Indigenous foods are a threat to their bottom line. Entertainment giants need the child of Malaysia to pray to mass produced toys rather than to play with puppets made in the image of local jungle funsters. Cowboys are told to pray to beer companies and truck manufacturers or forever be denied the opportunity to perform their life-threatening ballet in public.

All across the U.S., huge corporate interests are bidding to buy exclusive rights to serve their bilious processed food products to our children in public schools. For a few million dollars school boards hungry for funding are promising that their student bodies will buy only Pepsi or Coca-cola or ...? Generous on the part of the companies? Wake up. They'll make it all back and more by selling their horrible product without competition PLUS some of us will, out of allegiance to the good father corporation, buy only that same product. "Rule by undisputed sway" in the marketplace and legitimately!

"So where will the schools get that money?" I'm asked in a

plaintive tone. Our government wastes more than it would take to grant every child in this country a fine education. If we demand it. But we don't, we won't. We are hogs at the corporate trough too stupid in our stuffed daze to give a fig. Our public schools stink, they are proof positive that an abiding concern for the lowest common denominator is just that and nothing more. Any child can tell you that *Moral law* and *Individualism* are not video games. And community? It's so sweet and mysterious and illusive. Community...

Community: a. any group sharing interests or pursuits. b. an interacting population of different kinds of individuals (as species) constituting a society or association or simply an aggregation of mutually related individuals in a location. c. a group of people marked by common characteristic but living within a larger society that does not share that characteristic.

Individualism: a. the ethical doctrine or principle that the interests of the individual himself are or ought to be paramount in determination of conduct: conduct guided by principle. b. the conception that all values, rights, and duties originate in individuals and the community or social whole has no value or ethical significance not derived from its constituent individuals. c. the doctrine which holds that the chief end of society is the promotion of individual welfare and the chief end of moral law is the development of individual character. d. a theory or practise having primary regard for individual rights and esp. maintaining the political and economic independence of individual initiative actions and interests (as in industrial organization or in government)

I am often accused of being destructive. Of being someone who tears down. Someone unwilling to compromise. Someone who is anti-social, anti-progress, anti- just to be so. I recently received a letter from a Small Farmer's Journal reader I want to share. And I offer my response hopefully for some small clarification.

I believe in independence, in individualism, in community. I can see how some of my accusations may be felt as confusing attack on those who might reasonably be seen as members of a community that tries to welcome me, us.

Dear Folks,
Don't be too hard on environmentalists - think of the two sides' motivation:
* Environmentalists want to keep the earth resources from being exploited;
* The other side wants to trade it for money and/or power.
I consider myself an environmentalist but still love horses, small farms (organic), and freedom. If we are ignorant of facts, teach us, but don't attack our motives.
(signed) Iowa

Dear Iowa, You say there are just two sides to this equation.
There are the environmentalists -
and the other side.
I'm on the 'other side' but not by my choice.
I have worked my entire adult life striving to protect the earth's resources from being exploited. My time in the trenches is longer than I can remember, my scars are real, I will die engaged in the fight. I take pain at your insistence that because I do not choose to be a member of your "collective political party" (and that is what it has become) that I am on the "other side" - but it is not a new experience to be thought of this way.
Dear, dear Iowa, environmental organizations fight me, as a small organic farmer, rancher, tree-farmer, they fight me. They say to me I pollute the earth by applying manure to my fields,
While petroleum giants have succeeded in destroying the oceans.
They say my cows pass gas out their lovely back ends

which damage the Ozone,
 While the U.S, government stockpiles deadly nuclear waste and chemical weapons.
 They say that I am immoral for selling my cattle to slaughter for meat,
  fill in the blank.
 They say I subject my poor horses to involuntary servitude,
  fill in the blank.
 They say that my lands would be better for all of us as a wildlife preserve.
 But it isn't about me or mine. It is larger. I would suggest to you that there are many sides to the battle. There are environmental organizations, there are individuals, there is government, there is big business. In my humble opinion, government and big business are chummy with the environmental organizations because they have seen the need for political merger. And the vigilance of environmentalists needs a target, so the corporate attorneys and the federal granting agencies point out the little independent operators and say there is your villain "sic 'em."
 I'll work with you to keep the earth, but don't call me an environmentalist. I'll not sign on to the collective. I'm an individual first, just like you. We can work together as community if you wish. But know that I'll be riding my horse in search of all the lizard's hideouts. And along the way should I see agents of the lizard picking on my friends, I'll force them through a soul detector.
    signed arrogant

 I'm still angry.
 Even more so because I know it needn't be this way. Our lives are not linear. It isn't a two-sided world. It isn't a question of either / or. We don't start at point A and go to point B. Our lives are circular complications just as the bumper-car ripples on the surface of a rain splattered pond. We are everyone who came before us. We are part of the yet unborn. Our lives are not short

and tragic, they are long and wondrous as they reach forward and back and sideways within our own community.

So why do we cheapen ourselves by wasting this wonder of wonders, our most precious lives, in the lazy stupid consumption of a bland and unchallenging convenience. Why do we drink diet sodas and eat potato chips and watch television and hunger for motorhomes and winning lottery tickets? Why do we feed the lizard and pray to it. Why do we ride the fascist corporate choo choo to nowhere?

We, each of us, sense we might wake up one morning in absolute terror knowing our life has been wasted. Why can we, then, not walk away from consumption and into production? If we were to follow the little voice inside ourselves to creativity and thereby foresake convenience, I am certain pollution of the environment would cease. I am certain lives and communities would bloom.

But to do that we must get small, embrace insignificance, and learn to become farming tinkers. Farming isn't the answer. Farming is a handmade envelope within which might rest recipes for answers. And tinkers might just make the best farmers because they are craftspeople whose scale of concern and work is small and independent and lighthearted and light giving.

It is tinkers I want to talk about because we must talk about solutions. And tinkers is the solution. Little craftspeople, tinkers, bearded, aproned, jolly, mischievious, whispering in giggles, moving great mountains of impossibility in flutters of insignificant movement. Tinkers mad at the lizards and happy to be so. Tinkers who know together that the lizard's choo choo must be stopped.

Chapter Four

Wilbur's Farm
Potential, Influence and Realization

What a joy it is to have desire and goodness fit together plainly and to the benefit of all. I am thinking of our desire to farm and the great possible value, to the entire living world, when all our individual dreams are realized. And we are made more certain of this when we are exposed to evidence of denied agrarian dreams, squashed hope of self-reliance, and the limitless misery and fear of so many living in cities.

Farming teaches us, so completely, that wealth is in the interconnected "making." In this regard, the living thing or system with the greatest "potential" contains great wealth. And the living thing with the greatest "influence" creates great wealth. And the person who holds the "realization," (knowledge born of understood experience) of living, reproducing, natural systems holds the greatest wealth.

You want to farm. I want to farm. I am farming. You might be. If you aren't, you can. It is possible and it does not depend on your having a big lump of money to invest. Instead of money, you need "potential," "influence" and "realization." Money is no guarantee of access to, or success with, farming. There are rules which govern the hording of money and they often go counter to those natural laws which orchestrate the best farming adventures. How do biodiversity, sustainability, and interdependence measure against the profits, depreciations and capital gains of classic accounting?

"Yes, but how will I buy the land?" you quickly ask.

Slowly. And you will pay for it with potential, influence and realization. But you will never actually, completely, own it.

"Make sense," you reasonably demand.

You want to farm so why confuse the issue with possession? Find a place to farm, accept that your time there is a temporary wonder and forget about finding a farm to possess. The very best small farms are beautiful, fragile living collections of overlaid crystalline-like relationships which are easily altered or destroyed, in a moment, by a change of ownership, or tenant. The best farms are temporary; too often limited by the vision of one proprietary farmer.

In the early seventies I had occasion to drive along Interstate 5 commuting from a ranch I managed in Drain, Oregon to purchase supplies in Cottage Grove. On the west side of the freeway spread across a sidehill was a jewel of a small farm. Multi-colored, multi-supered, beehives punctuated the pert little pasture fields which encircled a warm and lovely, older, well-kept farm house. All visible from the freeway, the drive straight uphill to the house was lined with a riot of flowers in constant bloom through-out the growing season. A milk cow, a small flock of sheep and a donkey "possessed" the pastures and pond shore. For that entire summer, each freeway drive-by saw some new accomplishment or project underway, whether it was lumber being sawn on the small mill, or a new fence, or the little Allis Chalmers tractor mowing hay. One day late that summer I noticed a big hand-lettered sign advertising HONEY FOR SALE. That was my opportunity to meet the person who lived there. I had long thought that whoever it was certainly enjoyed a stroke of luck, or lots of money, to have such a lovely place.

Then I met Wilbur Long and my notions changed gradually, slightly. Wilbur was a tough, handsome, thin man in his sixties, of average height and with a springing step and a constant twinkling smile. He was a bachelor, living and working alone on his farm. Wilbur appreciated his farm life, good humor, hard work and pretty women.

I bought honey from him and we talked about my interest in learning beekeeping. Wilbur seemed the perfect picture of someone's appreciative, capable, cheerful hired farm hand. In this regard my notion changed and I came to wonder, if only briefly, if Wilbur was possessed by the farm rather than vice versa. Surely he had been there all of his life?

But then Wilbur told of how he was only recently retired from a life's work in a factory. Forced to retire early because of a crippling arthritis. He said that only a few years before our meeting he could not open or close his hands. They remained seized up in a claw-like position. Disabled, disgruntled and forced to retire a bitter man, he came across a magazine article extolling the curative properties of bee pollen and bee venom for arthritis conditions. The dramatic results of his own personal experiments with over-the-counter bee products led him to the "why-not-what-do-I-have-to-lose" conclusion that he would start life all over again as a beekeeper. So with therapeutic determination and very little money he set out to find a small farm to raise bees on.

He found what he described as a rundown house, a steep eroding 40 acre hillside which seemed to slide unwillingly right into the intrusive passing freeway. It was abandoned, unattractive and had been for sale, at a giveaway price, for quite some time. This was to become Wilbur's magical farm.

Wilbur set to work, clumsy at first we might imagine. And, he said, his first mistakes with bees resulted in occasional stings that brought the marvelous change of ever freer movement in his joints. Wilbur announced to me straight out and with the solid smile of eternal life-shaping gratitude "those bees cured my arthritis!" True or not, certainly the man I gazed upon stood erect, moved freely and exhibited the strength which is the gift of good health and good anticipations. So Wilbur proceeded to develop his little farm all with the goal of making life good for his bees. He planted acres of flowers of every type, careful to mix them to guarantee a constant carpet of bloom. He planted clovers. And he gathered geese, chickens and a few grazing animals for the fertility and balance they could bring to the fields and flower beds. The animals required buildings and fences, so he salvaged an old saw mill, taught himself to fix and use it, and cut lumber from his own trees. With each project, and every passing day, he felt better and his happiness and gratitude swelled. He quickly had farm surpluses to sell and barter and his vibrant attractive fairytale farm became a visual treasure to all of us fortunate to pass by.

With Wilbur's growing comfort and release from pain came new tangential interests. Wilbur made beehives from wood he had sawn on his mill. He also cut and cured great slabs of hardwood and found himself artfully crafting grandfather clocks from the material. It didn't take long for a market demand to develop and Wilbur had a waiting list of customers.

My first assessment of Wilbur's farm had been that he was fortunate to own such a place. My second assessment, or notion, was that Wilbur was possessed by, or a willing slave to, this lovely farm. But as I came to know more of the whole story, I discovered that Wilbur and his farm had created each other. Wilbur had come with potential and influence and he found realization. The weight of his potential resided in his determined fearless push to be working with bees and tending a small farm. His "potential" (and that included limited funds and experience offsetting absolute determination) purely started him out. And his "influence" carried him on as he affected everything by his values, his choices, his designs. When he decided to raise chickens in order to have their manure to fertilize the flower beds which fed his beloved bees, he influenced his whole farms' weave and vitality. The complexity of interrelationships grew and created unexpected living, fertilizing, chelating aspects.

Wilbur's actions were not guided by some hording instinct, they were treated naturally out of his growing love and gratitude for the work and the pleasured comfort it returned him. And his spirit, motivation and the curve of his step were all swollen by the discoveries of new beneficial connections on his farm. It was as if this mature man's entire life, up to the point of his farm tenure, had merely been time in some waiting room. To hear Wilbur talk, nothing mattered before this farm. Wilbur had come to the "realization" his efforts had created the living form that this place took and this farm had, in turn, created of Wilbur a new entirely different, self-reliant while dependent, human being. He was purposeful and glad.

I moved one hundred and fifty miles away and lost touch with Wilbur. Half a dozen years later I had occasion to drive south on I-5 and when the freeway made that anticipated gradual bend I was momentarily frightened. All evidence of Wilbur was gone. The hives, the flower beds, the little mill, the board and batten sheds, the pasture fences and gates, the animals - all gone. In their place the freshly sided and painted house had a small manicured lawn and a new connecting colored-metal shop/garage building. What once had been healthy hillside pasture, now looked like an eroded mud slide with sickly thin patches of grass that would never support livestock. The pond was an empty hole in the ground. The place was antiseptic, even anorexic.

I remember my first thought, "Wilbur's gone." I made inquiries in town but nobody seemed to know anything. Wilbur was gone and so was his farm. I don't know the end of the story, only the physical evidence and what it suggests to me. So I choose to back up to the picture I prefer of Wilbur's farm alive. Because it braces me and because I now know a secret which Wilbur's farm illustrates so well. It is a secret NOT because it is kept as such but rather because most people do not want to believe it or cannot see it.

The secret is this: the example, the model, of Wilbur's farm is attainable, we can all do it but only, thank goodness, if for the right reasons. And we can have an unlimited number of farms like Wilbur's because the more we have the better off everyone and everything will be.

But it is important to note that the example of Wilbur's farm, or more particularly of the idea of many farms like Wilbur's, is a threat to government, large corporations and many social scientists. It is a threat because the prevalent economic and political system would erode if droves of people were to abandon the "conveniences" of a consuming middle-class existence for the "labors" of a producing agrarian/craft existence.

Wilbur watched no television, drank no sodas, bought no new cars, shopped at no malls, visited no health clubs, gave to no corporate churches. This wasn't because he was opposed to these things and sought to make some statement through his actions and inaction. He wasn't boycotting. It was simply that these things didn't factor into his life. He had no need or time for them.

Wilbur was busy bartering and selling honey, homemade beehives, cut flowers, eggs, goslings, apples and handmade grandfather clocks he fashioned on rainy days and in the evenings. Wilbur labored and produced happily, and he punched NO one's clock. And his example is real and possible for any who would but try. Yes Wilbur was exceptional but don't confuse that by thinking his was an exception, an unattainable example. Remember, Wilbur didn't go into this with lots of money - Wilbur didn't purchase a ready-made storybook farm. Wilbur acquired a piece of ground no one else wanted and he allowed his determination to become a specific vision open to the influences of each day. The result was Wilbur's farm. A precious, if temporary, physical monument to that one man's spirit and the possibilities nature offers every artisan/farmer/craftsman.

We need a countryside dotted with millions of farms like Wilbur's. We need it because it would return our nations to true food security. We need it because it would give people hope and purpose to supplant the misery which now rains down on all as violence and despair. This countryside of small farms would give us back our safety and our familys. It would give us back the taste of good food and true health. And the beauty of it is that this countryside of small farms is so easily attainable. It doesn't require new laws or an expensive government program. All it takes is an ever increasing number of folks with the determination to be farming on a small scale. With each new success comes yet another proof it can be done. All it takes is "potential" (yours), "influence" (yours) and "realization" (yours and ours).

I never had the opportunity (or perhaps better put - I never took the opportunity) to thank Wilbur Long for showing me how to keep bees and for imprinting me so solidly, and so well, with his wonderful example. So I take the moment now; Wilbur wherever you are I know your liberated artisan hands are covered with gentle loving bees and flowers surround you. Thanks for showing a way.

Chapter Five

small tokens
up fiddle creek, then bear creek

Sources
> *sometimes artesian*
>> *once or more maybe explosive*
>>> *why not romantic*
>>>> *often*
>>>>> *only*
>>>> *small tokens,*
>>>> *embryonic*

lose to four decades ago, I managed a coastal Oregon cattle ranch; Angus/Jersey cross cattle on 670 acres strung out for miles along a narrow wet creek bottom behind Lake Siltcoos. This skinny valley was framed on either side by steep eroded coastal mountain sides covered by a dense three-tiered forest which made it unnecessary to fence the boundaries. Short cross fences, from mountain side to mountain side, were enough to keep the cows in place. Over a hundred inches of annual rainfall resulted in gargantuan forage crops, but the proximity to the ocean, with her pervasive creeping afternoon fogs, made haying somewhat of a challenge. We filled silos with dough-stage oat haylage and legume-rich grasses. The hundred and twenty-five mother cows ate well but the winter, pulling silage out of the concrete tower and spreading it by hand in the feeder to the penned cows, was intense. I remember thinking that for as much effort as I was putting into the winter care of these beef cattle, I should have been milking them.

This was dairy country. For better than a hundred years farmers had successfully taken advantage of the climate and forage to operate profitable small family dairies. Corn wasn't part of the indigenous mix but excellent clover haylage was. And this country loved a good Jersey cow. In fact the reverse was true as well. Creamers were still paying a premium for higher butter-fat content and the little Jersey cattle, queens of the rich milk, seemed put on this wet patch of earth to wander their tawny frames through swamps in quiet pursuit of secret milk enriching herbs, mosses, lichens, and legumes.

Along this small separate world of the Fiddle Creek drainage were a handful of farms and ranches. I was on the last place on the road, the turnaround for the mailman. Before me there were a handful of family dairies; one belonged to Harley Huff, a one-armed giant of a man whose physical size and stature were over-shadowed by his boundless energy and good nature. And then there were two brothers with their expensive Harvestore silos and largest, most modern of operations. Then came the first dairy on the route, that of Hugh Martin, prankster and comedian hiding a dark and tragic curse. Off a side shoot, a tributary of Fiddle Creek, called Bear Creek, lived Ed and Bertha Dahlman with their small and outstanding herd of beef cattle. Hugh and Harley and Ed were my friends, my role models, my heros.

At this point in my life, my early twenties, I invested a tremendous amount of energy in trying to convince myself and the world that I was a whole, complete, and well-formed adult with something to offer and a presence to reckon with. It was comic. I was a perfect dupe for the fun-loving practical jokes of my farmer neighbor friends who saw it as their God-given duty to apply the ointment of humility to my as yet partially formed character. All, that is, except Ed Dahlman. What Ed saw was a young man hungry for the shaping and deserving of simple

respect. I am sure that my having survived some truly life threatening practical jokes did help to shape my later personality, but I
am equally sure that what Ed showed me, explained to me and
illustrated with the example that was his and Bertha's whole life,
filled me with purpose and the beginnings of a vision.

Ed was a long, tall Finlander who retained his soft accent
and gentle northern demeanor. With his new bride, Bertha, he
had traveled around the west during the depression, in a Model
A, pulling a camp trailer. His young bachelor days had been spent
working as a buckaroo for large high desert cattle outfits. He had
been more enchanted with the potential hidden beneath the hard
realities of the existence than with any romantic assessment. He
learned to love the little details of his life and good cattle. He had
been excellent at his work.

I don't know when they had come to purchase their 600 odd
acres of coastal Oregon forest and bottomland. When I met them
they had already done a lifetime of work there. And the evidence
was everywhere to be seen. Ed was a blacksmith of the old crafts-
man order, and he was a logger, and operated his own small
sawmill, and built his buildings and fences, and raised a small
herd of the finest cattle I have ever in my lifetime known or seen.
And lest this sound lop-sided, it must be said that Bertha was Ed's
full-time working partner in all things. They had raised a son who
at the time was off in Alaska.

And they were artists. They gave me a new appreciation of
just what it could mean to be an artist. They didn't work in any
usual medium nor produce any customary art objects. What
made them artists was their vision. They saw, in every waking
moment, the structure of beauty in all things surrounding them.
And they intuitively, gently, played with that structure to create
an organic, living, six dimensional extended work of art of their
lives and ranch.

Ed gave me my first introduction to many things that would
later shape my idea of possibilities, beauty and ugliness. One such
introduction was the notion that a forest can be a perennial
garden orchard and that to view it as a resource to be mined was

to, as he put it, "pass gas in the face of God." Ed showed me how it is that greed becomes a debilitating force which denies that we ever become all that we can be.

If I recall right, the Dahlmans had about 150 acres of rich bottomland in pasture grasses and the balance was several hundred acres in what looked like an ancient untouched virgin forest of Douglas Fir, Cedars, Chinquipin, Yew, Hemlock, Maple, and Alder trees. Some of the Firs were more than six foot through at the butt and tall enough to give you a dizzy spell looking up from the bottom. The forest floor was shaded three times, once by the evergreen trees, again by the deciduous trees and wild Rhododendrons, and yet again by brush and bushes. And down under that last low canopy was a teeming world of activity producing a bumper crop each year of Chanterelle mushrooms. These woods supported deer, elk, black bear, mountain boomers, bobcats, eagles, hawks, owls, and the immeasurable millions of tiny living bits that created a living carpet spilling into Bear Creek to feed the trout, and salmon fingerlings gaining strength to make the metamorphological journey 12 miles to the Pacific Ocean. All connected, all magic. All beyond scientific interpretation, all beyond the ability of the urban environmental activist to grasp, appreciate, or truly honor. Because, as Ed showed me, there can be no true understanding nor honor of this world without complete immersion in it, without the tearful humility that comes of feeling so completely insignificant and helpless in that immersion. And yet recognizing that you must steal, rip, carve, from this secretive powerful magical world in order to have a chance of survival for yourself. Ed Dahlman showed me how it was that we could be so humbled and insignificant yet hold in our hands the seeds of total destruction for this fragile world. "It's beautiful," he would say, "but we have to go in and take a few trees out. We must harvest some even beyond what this forest gives up each year through windfall and disease. It's our part to do it. Because you see we, Bertha and I, are part of the living mix in this forest. What we do here is as important an animal impact as that of the birds, or fish, or other large mammals. So we choose to be discriminating and select trees which are not growing so fast, or crowding out others, or causing some other problem in the mix. And we take them out very carefully so as not to damage the remaining trees." The result was that the Dahlman forest was healthier, stronger, and more productive than neighboring tracts of public forest lands protected from logging. As Ed would say, "the test is, have we helped this forest by our being here?" And Ed and Bertha passed that test because though they took timber out each year for 30 plus years, to make into lumber and sell, there was more merchantable timber remaining than when they first arrived there. And they passed the test because beyond counts of dollar value they improved the overall health of that forest increasing its fertility at every level.

And the temptation to be greedy and take more, take it all, never entered the picture. It just didn't exist.

In later years when I was involved in progressive causes that included early organizational efforts to halt clear cutting of timberlands, I approached Ed and asked if he would allow his operation to be used as a model of sustainable timber practices. He refused and seemed hurt that I, of all people, didn't understand that to be a violation of who he was. How could he set what he did up as an example of what was right when from day to day he wasn't sure himself. He was only doing what he felt must be done in his small corner of the universe. No, he wouldn't allow it because it would only lead to arrogance. "Anyway," he said, "it would be greedy."

The Dahlman Bear Creek Ranch could have run a hundred to a hundred and fifty head of mother cows by all usual measures. And that was with putting up a full compliment of winter feed. But Ed and Bertha ran only 30 to 35 head of cows, because they believed it was all that should be run. I was a hot-shot young rancher who equated fully utilized carrying capacity with the best opportunity for success. After all, wasn't the name of the game to get the most money? Didn't that mean you had to get as many pounds to market as possible? Wasn't that the criteria for measuring success? The Dahlman example went counter to all that I had been led to believe by prevalent industrial models of beef cattle management. Ed redefined for me the difference between intensive and extensive agriculture.

With my first visits to the Dahlman spread, I came wanting to see a successful small ranch belonging to and being run by a man who had formerly worked for others as a cowhand. I wanted that man to be someone who had come from a background of limited financial resources. I wanted that man to be self-taught and self-actualized. I was looking for a mirror-image of myself projected into the future and loaded with my notions of success. I wanted to see, by Ed's example, that I was not only on the right track but that I also had a good chance of making my dream come true. He was to be proof of that. Subconsciously I had gone much further, into prejudice and close-mindedness. If I had found failure in Ed's example, every pore of my being was posed to point rapidly to the failure as evidence that the plan/design of operation differed substantially from my own and it was for this reason that it did not succeed. I was not prepared to be shown other outlines; outlines for success which depended primarily on a deeply held honor of the mysteries of nature's design and a radically humble notion of what success itself meant. I was young but I had begun to form a shell. Ed Dahlman deftly peeled the shell away and set me raw in the middle of a plate of unanswerable questions and gently poured over me a sauce of possibilities.

On my first visit, after handshakes and a quick, albeit mistaken measure of the man, my view was drawn to the adjoining pasture and ten steers. I was a student of beef cattle husbandry and prided my young arrogant self with knowing 95 percent of the world's cattle breeds on sight. It came from looking over literature, not from walking pastures. I turned to "Mr." Dahlman and said, "Are those Blonde d'Aquataine cattle?"

"I don't have a clue what you just asked me," he responded.

"Those big steers, what breeding are they?" I asked.

"My own," he answered with a wry smile.

Not wanting to seem impertinent I said, "The color and size and shape of these cattle, they aren't Charolais, I thought maybe they were of another French breed called Blonde d'Aquataine."

Not missing a heartbeat he said, "I started with a handful of the best Shorthorn cows I could find, all of them white. Bred those to the best Charolais bull. Then to a Hereford and back to white Shorthorns. That was a long time ago. I keep the families separate and raise my own bulls now."

"How old are those steers?" I asked thinking they seemed quite large.

"Two year olds. They'll be going to the market soon."

These steers were beautiful, long backed, well-proportioned animals of a rich honey-cream color. They were big cattle but neither long-legged nor stocky. I would have guessed them at 1,400 pounds each and prime.

"You been graining them?" I asked.

"No sir, they are lean and grass fat," he said.

Ed's steers were both impressive and confusing to me. Most mother cow operations out west raised a calf crop they sold each year. These calves went on to other pastures as feeders and eventually to a feedlot for finishing with high-octane supplements. I knew Ed's operation to be cow/calf so I was surprised to see the ten big two year old steers.

Later, at a different pasture I came to see his little cow herd for the first time. I was thunderstruck. These were the most attractive beef cows I had ever seen. Along with the obvious fact that they were all that same honey-cream color, to any experienced cowman's eye these animals were absolutely identical; same bone structure, same size, same heads, same top line, same girth, everything the same. Here were thirty of the prettiest mother cows I had ever laid eyes on, printed page or pasture, and by their sides were thirty big strapping calves of equal uniformity. Dahlman succeeded, in the weeks that followed, to teach me that to look across that rich green pasture, bordered as it was by Ed's delicious virgin-appearing forest, was to view a true "breed" of cattle in "its" setting. This was his life's work, his art.

Perhaps it was something in his attitude towards me, or maybe the whole picture was just too much, but I quickly began to melt. Bits and pieces of this picture weren't fitting my expectations yet the whole view was enthralling me, changing me. To my eternal credit, this breakdown in my assuredness allowed that I "knew" this to be one of those personal discoveries that would come to shape me or push me or both.

Ed woke me up with a question.

"Which of those bull calves would you select as a herd sire?"

The question catapulted me back into my former arrogance. Looking at the pairs grazing, I saw that some calves were laying down, some were grazing or nursing, but one proud bull calf was standing stoically as if to say 'I may only be three months old but I am somebody.' I took a minute to measure him up, only a minute, he seemed perfectly correct - a real eye pleaser with good size. I pointed to him.

"He's a cull," Ed said immediately as he casually pointed in the other direction at another bull calf that was busy grazing. "That's the keeper. Do you know why?" I looked at the calf. He was identical to the one I picked out except for the fact that he was busy eating. He didn't have that certain 'I am a bull' look or presence. In fact he seemed just a bit clumsy as he stumbled forward. It was as if he couldn't eat fast enough. Before he finished with one mouthful of grass, his eyes caught another and his feet started pushing to the next spot. Eyes and feet forward but mouth still busy back here. He looked like he expected a bell to ring sounding the end of lunch and he was still hungry.

Saving me the embarrassment of my answer, Ed offered, "it's because he's always eating, can't seem to get enough. He'll breed that into his calves. What would you rather have, a good looking bull or calves that eat and grow fast? It's not enough to be observant, you have to understand or respect what you're looking at."

Ed raised replacement heifers, his own bulls, and kept the steer calves to finish on grass and sell as two year olds. He mastered each step of that husbandry process to perfection yet was always looking to improve. When it came time to sell his steers Ed would call the Portland, Oregon livestock auction yard, some four and a half hours away, and tell them he was bringing in six to ten head of his steers. He only took his culls to the nearby markets, the good stuff went to the city. And there was a reason. Portland treated Ed right. Over the years, Ed's steers had developed such a devoted market following that buyers would be called and make arrangements to be at the yard for Ed's handful of steers. And at the beginning of the sale the auctioneer would announce that later this afternoon we'll be selling a set of Dahlman steers. When the animals did finally enter the sale ring the auctioneer would say a few brief words about how special these animals were - but those words were drowned out by the deafening silence of all those stock people sucking in their breaths as they wondered at the perfection of the beasts. The animals always brought 200 percent or greater of the current market price. And after they sold, you'd hear a rare burst of applause.

Art, all of it was art.

The artistry didn't start at one point, nor end at another, it was everywhere on the Dahlman spread. For example, Ed didn't put up winter feed like other folks did, he thought such practices were wrong, bad for the soil and the plant health. Instead Ed would take his Ford 8N tractor and clip his pastures, relatively high, and rake and bale these clippings to put a small emergency supply of feed in the barn. His goal was to manage his grazing so that, in the open wet winters of coastal Oregon, his cattle were always on pasture and never needing put-up feed. The year I met him he had three years worth of clippings stored in the big barn because he hadn't needed to use them. But even so he would, nevertheless, always clip those pastures not because putting up hay was important but because clipping the grazed grasses and legumes stimulated more growth and in his opinion improved fertility.

Often, when we'd walk his pastures, Ed would make me get down on my knees. He'd take his big pocketknife and cut a wedge of his pasture and show me the root system. He'd say, "I don't know everything about these roots but I have figured out that they go through changes that can help the dirt. Clipping the ungrazed grasses and clovers before they quit growing seems to release goodies into the dirt that make them grow that much better. I'm sure you, with all your book learning, will think I'm crazy but I know it's true - I've watched it for years."

And Ed's pastures, to the modern agriculturalist, may have appeared to be over-run with weeds. But Ed was fond of saying, "There's no such thing as a weed, that's just a word people came up with to describe an inconvenient plant. These so-called weeds sprout up because the natural balance has been messed up. The weeds are there to try to correct the balance. You notice how the Alder trees come back in thick when someone has clear-cut a forest? If you take a few trees out selectively, the Alders don't rush in, why is that? Now, Alder's not a bad tree, it's just inconvenient if your job is to grow Douglas Fir. But the Alder has a place, a job, we just might never know exactly what that job is. I think we should respect all the plants of the forest and that includes these pasture plants."

I must not have looked fully convinced because Ed made me follow him to the back end of his ranch and a narrow little fjord-like pasture. In there he got down on his knees and summoned me to follow. There he pointed to a small delicate yellow flower. Once I saw it, I noticed the plants that spawned this flower formed a sporadic undercarpet in this little pasture.

"It only grows here," Ed offered, "no where else on this 600 acres will you find that plant - just here on this five acres."

Still not picking up on the import of his words he added, "You know, I've told you it's not enough to be observant, you have to understand or respect what you're looking at. Many years ago I was watching my cows, had them shut in the forward pasture, they were belly deep in clover and grass and should have been content but they weren't. I watched as one cow jumped the fence, knocking down the top rail. I went and got Bertha and we opened the gate to herd her back in but she was gone heading somewhere up the valley. And in no time all the cows and calves streamed out the open gate following the renegade. I swore I'd sell that jumping cow as we got in the Jeep and followed the cattle. They all came to this little canyon floor we're standing in and it was Bertha who noticed that they were all happy eating. 'Eating what' I said. 'Those little yellow flowers!' she said. Sure enough, that well-fed herd had traveled a mile and a half with the single-minded mission of eating some of these little yellow flowers. Why? I don't know. Never will know. What I do know is that we have happy, healthy cows, pastures, and trees maybe because we leave the little mysteries alone. When I don't understand what I'm looking at, I make it a point to respect it. After all that's how you and I came to be friends." And he smiled a Finnish smile.

The Dahlman's were enchanting to me. They helped to teach me how to doubt my prejudices and love the little details of this life. They taught me to respect those mysterious small natural tokens we can never hope to understand. The only way I can begin to thank them is to try to share their poetry. LRM

Chapter Six

Civility, Frugality and Independence
The Orange Machines

I was at a family reunion recently. The occasion was the fiftieth wedding anniversary for my folks, Ralph and Lydia Miller. All five of their offspring, four sons and a daughter, were present as well as many spouses and grandchildren. We are a scattered family living from Oregon to Florida to Alabama. Getting together had never been easy. Because of our parents' anniversary celebration and our far flung separateness, the experience had an almost painful urgency and clarity. Some of us laughed too hard, cried too easily, and said things without pause. All of that is of course as it should be. And the reunion sprouted the seed of a recollection from childhood which I would like to share.

During the reunion I was surprised to feel myself flinch each time I was playfully acknowledged as the sentimentalist of the family, the romantic of the family, the hillbilly of the family, the hermit of the family, and on and on. I could not name the major league teams of any sport and (unthinkable!) I was completely in the dark as to the personalities of professional golf. I was not a member of any political party and subscribed to no urgent post-modern theology. In my feeble attempt to fit in, I made lame attempts at contemporary jokes about Clinton and Gingerich. I didn't quite fit the mix. It caused me to pause and wonder how it is that a big family might go in so many diverse directions and come to hold such different values. I struggled to remember some shared experience of our growing up that might have given a clue to why we had become so different. I found a memory, in that search, that has told me other things. I found no answer to my first question but I have a suspicion. Without getting into any potentially embarrassing personal specifics about my sister and brothers, I'll just say that the diversity in our family is a clear evidence of health and fertility. Somehow my parents, and the circumstances of our growing up, instilled in each of us a priority for self reliance. That stuck and it scattered us. They tried, as did the fifties, to teach us tolerance and thriftiness. It is still too early to know if that stuck. I want to believe that, measured today, each of us now middle-aged Miller kids, have become good citizens. Certainly the only questionable one would be me, and that because my neo-luddite, anti-social, pro-farmer tendencies do not measure well inside certain corporate definitions of citizenship.

Back in the fifties my father purchased a large lot in the middle of an orange grove on the outskirts of Fullerton, California. He proceeded, with some questionable help from his four small sons, to build a unique and wonderful home for us. Then followed the great southern California building boom and the farmland and orchards receded rapidly as scores of tract homes popped up overnight. The clash of the orchards and the construction provided us with the backdrop, and often the materials of our growing up. In that setting there was one insignificant but very unique, specific, and peculiar evolving and repeated experience which recently found its way back into my memory banks.

In the evenings, after the builders had all gone home, we kids would form an insect-like wave of little creatures scurrying to check out the carpenter's and mason's accomplishments for the day. We weren't supposed to go anywhere near, it was too hazard-

ous. But the only way you could keep an entire neighborhood of
adolescents from canvassing and claiming such a stupendous
playground would have been to offer an alternative (TV had not
become a magnet yet) or physically restrain us (the adults were
too tired in the late afternoons to successfully link bowlines and
handcuffs). I don't know how it started but I do have vague
recollections of our adventurous gymnastic attacks upon the
mason's big piles of damp sand, dumped at the sites for addition
to cement mixes. We would run to the top of them and then
plummet or roll off the other side in a roar of giggles and pubes-
cent expletives. My siblings and I, along with whichever neighboring kids we were partnered up with at the time, would "kind
of" stick together. I am the oldest so I was forced to recognize the
real substantive threat to future liberties implied by my parents
insistence that I was somehow responsible for my younger brothers and sister. If they didn't come home with me each evening (or
before me) and could not be completely accounted for during our
"free time," not only was I in big trouble but WE would create
enough anger and fear in our folks that they would take the time
to restrain us from future evening sorties. So, by necessity, I had
to resort to creative ways to keep us together. This I did by telling
horrific stories about what might happen if any one of us should
lose sight of any or all of the rest of us. Images of deep holes
hidden under piles of boards, cement mixers that came to life at
night and ate children to fuel themselves for the next day's work,
and exposed nails dipped in poison were used to paint a picture
of the necessity of "togetherness." Often I was too effective and
one of the youngest would run home crying which meant we all
had to follow him or her and any additional evening's adventure
was cut short.

I can't take credit for what happened, but it was perfect. It
was like a scene from the "Little Rascals." The evolution escapes
my memory - I race forward to mid-process. Now picture this, we
go straight to the building sites anxiously looking for new freshly-
dumped piles of wet sand because they formed symmetrical
volcano-shaped piles five, six or seven feet tall (big by kid stan-

dards). Then we'd comb the nearby area for oranges, preferably small ones. With a dozen or so oranges we'd return to the pile and begin a furious transforming sculpture process. Beginning at the top of the pile we'd flatten a staging area or platform to set our oranges on. Next funnels or slides, shaped to fit the size of the oranges, would be patted into the side of the pile running straight-down and sideways and sometimes up just a little. We were in pursuit of a certain physical truth that was measured quite simply by whether or not our creation worked. We wanted to be able to push an orange off the top platform and have it follow a path down, sometimes fast, sometimes achingly slow, sideways, up over jumps, under bridges, through tunnels, and ultimately to the bottom. In the beginning our efforts were pretty basic and laced with arguments. So we'd stake out claims to this or that side of the pile for "MY" slide area. Then someone would send his or her orange down the pike only to be surprised by some increased speed or weak funnel wall to have the orange "escape" and take off down a neighboring path. Yikes! That was fun! Soon we'd deliberately start setting up "possible" and "probable" ways to connect this run to that run. The result was that we created many truly incredible "orange machines" that could only be appreciated if you were to first watch the enthusiastic cooperative efforts of these several independent kids - and then could stand back and watch as dozens of oranges were launched to race, simultaneously down and around. Sometimes narrowly missing each other, at times crashing into one another, and often surprising the sculptor/architects with result and consequence. Each night we'd go home feeling full of ourselves and each other. Next evening we'd race back to the previous day's handiwork fearing the worst, that the pile we'd shaped had gone into the cement mixer. If it hadn't, we'd apply our ideas for improvement and enjoy the orange machine for another few hours. If it had been destroyed, we would run off to other house sites, scanning the horizons for new candidate piles. I don't know how many of these sites we created. If I believe my memory's wishful manipulations, we made hundreds over the course of years. But my memory does like to swell on these good points and it tends to distort to advantage. Just as I know that I never was a great teen-aged surfer, I know also that our orange machines must have been product of a short season or two. But I do clearly recall that we got good. At one point we got so good, and justifiably proud, that efforts were made to get our parents to come and see what we'd created. I don't think they ever did. And the adventures of the orange machines, having successfully baby-sat and entertained and schooled us through a dangerous and potentially boring piece of our lives faded into games of tag-football, sandlot baseball and homemade coaster cars.

The citizen of highest value to his or her community often embodies the qualities of tolerance, thriftiness and self-reliance. Though few would want to admit it openly, many of our educators, politicians, church leaders, and the officers of large corporations are aggressively opposed to such qualities in their respective charges and constituencies. Tolerant, thrifty, self-reliant people are difficult to herd, frighten, fleece, numb, or confuse. Certainly we orange machine builders weren't to be numbed or confused.

A tolerant manner is foundation for an individual's civility. And when a people come to easily treat one another in a civil fashion, society gains. Simple enough and good. It is far better than what we have come to consider as the norm for today. Today we worry after our destructive and uncomfortable society. It is easy to point, daily, to the glaring absence of simple civility within our modern society. Many of us treat one another poorly. We take family members and old friends for granted. We treat new friends with suspicion. We approach strangers, if at all, as if they were carriers of the plague. We've lost the ability to make our different "runs" work together.

Years ago, when I lived in San Francisco, I knew a man, Michael Donnelly, who made it a point to smile and say hello to every person he had occasion to face. In a pulsing diatomaceous city like San Francisco this was quite a challenge. Michael once pushed me to try it for a day. He said, "You can't imagine how good you feel just by getting an honest smile of thanks and acknowledgment from someone who just seconds before was closed up to everybody." He would argue that if we all were to follow that civil adage "speak only when spoken to" that the result would be a suspicious silence. Michael made of civility a personal form of poetic interaction with society and it fed his psyche well.

I believe that thriftiness or frugality makes, of each individual, a nesting place for appreciation. How can we take each meal, each laugh, each sound, each touch as full of the best measure of living when we spend our lives as though they were easy-come easy-go? Our lives are precious, hard-won and lost only in great tragedy and sacrifice. There is a great lesson in the example of those individuals who, upon learning that they have but a short time to live, discover heightened beauty and value in everything. In truth, it goes without saying that each of us has only a short time to live. To approach our time deliberately and slowly, savoring every aspect of each moment, is to actually become frugal and thrifty because waste and haste are almost impossible in such a frame of mind and spirit.

Being small farmers means being self-reliant folk. We not only know what it means to depend on ourselves, we also feel that knowledge forever tempered by the humility that comes of being constantly reminded of our true place in the natural scheme of things. We are small and dispensable and we are important and vital. Self-reliance lies within what seems to be that contradiction. It's actually a balancing of juicy and frightening circumstance and attitude.

I do hope that those of us with children will give some consideration to how the young ones spend their time and perhaps include a moment to marvel at what those children see as their accomplishments. The small children and young adults of today carry the seeds of abundance and clarity and balance and thrift and tolerance and self-reliance...

Chapter Seven

The Pirate Genius and the Public Genesis

As I sat on the curb one day
as quiet as I could be
A great big ugly man came up
and tied his horse to me.

    *The **Pirates** of old were those scoundrels who, with the aid of freed outlaws and devilish humor did ravage and plunder the ships of tyrants.*
    __Genius__ is that mix of creativity and swollen dormant seed ready to burst into new life.
    __Public__ is the commonwealth of many individuals.
    __Genesis__ is the process of beginning, the color of new, the promise of original thankfulness.

Farming is about connections. We aren't speaking of those important people you might know. We speak of how every action taken has a multitude of effects spreading like a net of interlocking ripples and extending far beyond our capacity for understanding or absolute knowledge. Our world is the sum of all that we do and don't do. No where do I see that so much as in farming. It is elsewhere, it is everywhere.

Science is so often baffled today, not by its questionable measurements of the physical world but by beginnnings. How did this or that start? It is almost as if 'they' think all will not be right with the world until we can pinpoint the origins of every-thing.

It flies in the face of science but, as most every thinking farmer realizes, every beginning is but a connecting tissue coming from somewhere and heading elsewhere.

When I was eleven or twelve, my father, a cement contractor and builder, took me on the job with him. I was to be paid to work. I remember how excited I was, how eager to prove I could do the work, how anxious I was to be in the same working environment with my bigger-than-life hero father.

When we got to the work site, he pointed to a foundation ditch and handed me a shovel.

"Son I want you to clean the loose dirt out of that ditch," he instructed.

"But Dad," I argued, "I want to work with you. I want to build things with you."

"This is what building's all about, son. First we need the ditch clean so that we can build a form for the concrete."

I was disappointed. This was definitely not how I had imagined my day working with Dad. I climbed down into the ditch and 'toyed' with the shovel and dirt. Some time passed and my father came back to where I had been "working." He shook his head, told me to get out of the ditch, and went in himself. I watched as he worked methodically shoveling the dirt and shap-ing the ditch sides and floor. I felt an itch. He must have seen that because he gave me back the shovel and told me to try again. I did and it went pretty good for a while but, as with any young-ster, I soon tired of the work routine and returned to my slow pace. Again he came to visit the site and said to me,

"It's not supposed to be fun. It's work. Try not to think about how much you've already done. Or of how much is left to do. Think about something else. Set yourself into a rythmn of digging and quit fretting about the work. Think about something else. We need this done, son, the cement trucks will be here tomorrow and we have a lot of work to do to get the forms up."

From that moment, from that beginning, I learned to work. The tedious work that can only be done one step at a time and sometimes with one step backwards. That moment has affected my entire life, spreading its ripple net in every direction. But that moment came from somewhere, it was not born in a void. My father took me to that site and he brought with us his values and personality. His gentle push and important example don't constitute a great story or powerful metaphor. They serve as a signpost suggesting where a beginning came from.

Connections, connecting tissues, beginnings - these are, at an ideological level, the fabric of life.

We are well into the early days of an age of 'genetic manipulation'. And this age threatens that fabric of life as it has never been threatened before.

Large corporations, and the corporate ethic which drives and sustains them, poison us, demean us and fuel our inhumanity. If corporations and their products and services were ignored, they would disappear as the dragons of old once did.

Recent innovations and global trends in corporate agriculture have presented us all with a new extreme urgency which screams for action on every front if we are to stop the slide into a corporate engendered and orchestrated death of many life forms and a resulting diseased and vulnerable humanity and planet. It's that serious. It's right now.

We have taken for granted that seeds could be planted in the earth's fragile topsoil, watered, tended, and the result would be food for us and our animals, green manures to return to the soil, and seeds which could be gathered - saved - replanted. The very economy and spirituality of that which we know as farming stems from this powerful natural simple contract of allowance and possibility. A contract which we hold with the biological world. A contract which implies a standard of conduct and stewardship on our part. A standard which, collectively, so called developed man is now at odds with. Modern man in his arrogance has violated his contract with nature. We are in default.

It is in man's nature to want to understand - to know more - to try to find where processes may be enhanced or improved. So with farming's simple contract we have always tried 'to do better'. Individually this has resulted in marvelous advances, not in nature, but in man's working relationship with nature. It has been a good partnership, that one between individual humans and the biological world, because a natural balance has tempered inquiry. It is, and has been, in individual man's nature to stand in awe and wonder of the majesty, mystery, complexity and power of the biological world, God's creation, our nest, this mind-boggling beautiful world.

Elsewheres, where individuality is a curse and an inconvenience to the constructs of collective profit, this balance has disappeared. Within the modern corporate ethic and prospectus, none of this manifest humility is real unless used to achieve greater profitability. And profitabilty sits on a mound of urgency and simple linear progression. "We must have profitability NOW in real dollars and cents."

(With modern computer technology and the Web, today's corporations measure profit trends HOURLY and are prepared to react immediately to keep profits up. Nothing else is more important to them. Nothing. Not even life.)

So man's curiousity and need to know, formalized into questionable scientific legitimacy, has come to require facilities, equipment, materials, funding. And the corporate boardrooms have identified that within molecular science and genetic research lies not only vast fortunes (many billions of dollars) but to the winners of *'the race to mutate'* goes nearly complete control of food, fiber and fuel for the planet's population. So corporations fund science. Science in turn serves corporations.

They prefer to call themselves Bio-tech companies. The ones

I speak of are those large transnational corporations who control
genetic research and development and rush to market the results.
I prefer to think of them as the *Frankenstein-factor.*

These Bio-tech corporations have declared WAR, biological
war, on farmers, farm security, soil health and cultural diversity.
They are moving, in the open, to control food production
throughout the world. They have displayed, in some cases, nearly
a century of total disregard for human life, and the planet's
ecosystem. They have made of themselves the enemy in the
darkest, simplest, most absolute and dangerous way possible. If
small farmers and rural communities, and perhaps even the
human species, are to continue with any semblance of health it
should be a goal to take these Bio-tech firms down. Barring the
interesting if remote possibility that this dragon, the Frankenstein-factor, might be dealt with in the courts (to date dozens of
courtroom losses with hundreds of millions of dollars in fines
haven't slowed them) it seems likely there are only two arenas
open that might account for their demise; public outcry and new
benign forms of bloodless guerilla warfare.

If the public in the US, Latin America and Canada and Asia
and Africa were to echo the anger of Europeans forcing local
governments to at least accurately label genetically mutated
products (with warnings) the dragon would disappear by lack of
profit. And as for guerilla warfare, it is a reaction I fear, not one I
encourage. And I fear it because rationales for appropriate
measured destruction so often adopt the adversary's ethics as rules
of engagement My personal fear does nothing to change the
urgency. Change is on the way.

In my travels and talks and meetings I see the germination
of a new deliberation towards change. Folks are determined to
stop the monsters. They speak **NOT** of guns and bombs, they
speak of warfare as a cultural activism with civil disobedience,
political sabotage and chemical interference as the tools. They
speak of the pride of wholesale secret membership in a loosely
organized and untrackable effort at global humiliation, market

avoidance, and pesky pranks all directed at the Bio-tech firms which constitute the Frankenstein-factor. They speak of changing into uniforms in corporate closets and running down the hallways masked and spreading leaflets to the corporate labor force. They speak of lawyers in holistic and spiritual servitude to the cause of stopping Bio-tech firms in the Patent courts, and in the trade courts, and on the golf courses. They speak of Baseball and Basketball stars and Movie stars paying for and appearing on TV public interest spots denouncing genetically engineered food. They speak of sneaking transgenic foods laced with roundup and rBGH and BT onto corporate banquet tables accompanied by warning labels which read *"proceed with caution, we produced this crap!"*. They speak of mass public seed gatherings and planting where individuals dare to "pirate" seeds. They talk about clever collective courage. They push a new form of modern piracy.

There are unconfirmable whispers floating out across the great North American plains of a secret dark order of farmer pirates, bound by composting oath, who, with willful disregard for the purchased laws of babykilling corporations, do vow to sow the seeds of their fathers.

They would tell you this is a cause worth giving our all to. They would tell you that all Bio-tech corporations in consort may already be the most destructive entity and/or agency to have ever existed. And they are right.

You might consider the cigarette manufacturers to be evil and destructive but they pale by comparison to what the Frankenstein-factor has already done, is doing and is planning. I am telling you that most Bio-tech firms display a willful disregard for life through corporate edicts and their murderous insidious technologies. They are a DIRECT threat to the safety and well being of our children and grandchildren and great grandchildren. They threaten to destroy biological life on this planet. So what is to be done?

A call to action. A Profit-Crimes Tribunal

Transnational or multinational corporations which display a willful disregard for plant, animal and human life (as well as ecosystems and biological communities) must be held accountable for their actions and the consequences of their actions. There should be formed a parallel to the *War-Crimes Tribunal* in the Hague. I propose a **profit crimes tribunal** wherein indictments might be placed of corporations and individuals (with legalized limited liability) charged with profit-motivated crimes against life. If any **corporation** should be found guilty of destroying any life form in the name of profit it should be executed, terminated, liquidated, cancelled, pulled down, run out of town, denied, zeroed, abandoned, folded, electrocuted, hung, gassed, and all of this without any return to investors - period.

Corporations are formed to avoid *liability* for their actions. This should not extend to a demonstrated willful disregard for life. To date, the only thing which may shut down a corporation from without is a fall in profits. It is high time that these monstrous legal entities be held more directly accountable at least for their crimes against life.

As I was thinking about this essay I was working to clean a 240 foot ditch to set up running water to our little ranch house. It was this setting that brought back to mind my childhood experience in the ditch. It was a hard job cleaning that ditch but these editorial thoughts took me elsewhere. In fact when I was almost done I felt a ping of regret because I was far from done with my thoughts about this difficult subject. I actually wanted more ditch to dig.

Beginnings? Perhaps there is no such thing. Or perhaps we're not to know. So what does the ditch and beginnings have to do with genetic engineering or piracy or genius? The ditch is a metaphor for a lot of things: the value and place of simple hard work - a way to get from one point to another - the birth of a sensibility - creativity -the future. Genetic engineering is seen as a way to get some place fast and that place is a more perfect and profitable world - at least in the arrogant greedy eye of the corporations. And piracy in some new sweet/sour design is what any of us, who dare to confront the corporate menace, are charged with. Genius is what I believe we must have if we are to be good pirates. And I don't mean the simple high IQ junk genius, I'm referring to genius as a sweltering, steamy, individual home to seeds of ideas struggling to be born. Maybe I mean the genes in us. Maybe I mean the genesis of us. Yes I mean us, as individuals, fighting to protect our genes, our seeds, our beginnings, our creativity, our humility, our futures, our children's futures and the ditches where they'll be discovering their sensibilities, our connectedness, our commonwealth. Our public genesis. Yes. It will take the Pirate Genius to save our seeds, our food, our fiber, our family, our future, our public genesis. And we are the Pirate Genius.

Chapter Eight

Fresh Fruits

It was in October '97, my baby daughter's third birthday. I was driving my pickup truck home to the ranch from the Journal office. A dark early evening.

I was startled out of my motoring reverie by the sight of a cougar lying in the road. She was spread out on the right side at the gentle curve just before first entering the ranch property's southern entrance.

She got up slow enough for me to see it was a large female with pronounced black face markings. Odd. They are usually such quick and illusive creatures, seldom seen. This one turned in the road and went slowly up the hill towards the rimrock. My first thought; maybe it was hurt. A wounded cougar can be lethal. But she moved with too much assurance, almost as if she didn't see the approaching headlights, or care about them. As she disappeared, I realized I had stopped the truck and that my heart was racing. What a magnificent animal! I then sped up to see if she was still visible from the spot she originally laid on. I stopped, rolled down the window and listened, my heart still pounding. The excitement was all in the sighting.

We live with the cougars out here, we know they follow the Mule deer migration. And now the deer herds are coming down from the Cascade Mountain range. But we never see the cougar, never even hear them.

I got out of the truck and with flashlight in hand looked to see if I could read the sign of the cougar's rest and crossing. The road bed is a hard-packed sand and I found no sign, none. Made me wonder for a split second if I'd really seen her but a quick check of my pulse confirmed it. Yep. And it didn't take long for me to realize that the experience had been good for me. Woke me up, shook me up, widened my eyes. It had refreshed me, enlivened me, nerved me, charged me. I couldn't wait to tell my family.

Before getting back in the pickup I looked around. We Millers live in a wild country of some unique character. With hundreds of thousands of acres of sparsely timbered brush covered rangeland, we have not been able to concentrate attention on each corner. At the moment just after the cougar sighting, I took time to assess this spot. I hadn't given the spot much consideration before.

Old hidden homestead site on the northwest side, with remnants of a hand-laid stone wall and pieces of a tree-speared ancient truck. Juniper thicket and tall grandfather sage brush mixed with Bitterbrush. Might deceive the anxious cyber-intoxicated extrovert to think it was a "nothing" spot. The dirt road cuts through this narrow canyon on the southeast side almost as if that haunted homestead needed skirting, avoiding. The sensitive self-ordained folk might wish to flee this vortex. Vortex? Feels that way. But why, what is it? Higher up you can see its canyon nature with opposing rim-rocked bluffs. Then I see the narrow gentle washed-out slide area to the south, a perfect way down off that cliff edge. And on the other side I look hard and finally pick out where the rim rock has accomodated with a brushy set of stone steps. It is a tunnel to the other side! Animals have learned that they can walk down those stones with some cover from view. They can then rest on the canyon floor, deep in the tunnel, amidst the thick brush cover and Junipers. Drink, if they would, from the springs hidden there. And then when the time is right, skirt that stone wall and dart across the dirt road and up the washout to the world atop that flat overlooking our ranch to the north and Squaw creek to the south. Naturally the cougar would choose such a spot, great hunting. Back a hundred years and more, before Fremont came through, Payuse natives doubtless waited in the thicket for game. And here I am parked in the road, with a flashlight and a book deadline: the keeper of transparent and doubtful urgencies.

I left the spot reaching inside myself for that rush of feeling, back when I saw the cougar.

It's past now and fading, the imprint of that moment. But it still means something to me.

I was glad to see it was a female. Last spring the state trapper had been out hunting for a large old male cougar which had eaten several lambs. Paul Reuter runs two large bands of sheep out in our neck of the woods. He employs two Peruvian herders who live with the sheep, their saddle horses and guard dogs. Up on the Geneva flats, past the old ghost town site, the cougar had managed to pick off a few stray lambs leaving ample sign to aid the trapper. Last I heard the cat had eluded all efforts to snare him.

The conventional wisdom has it that once a cougar starts picking off livestock they get a taste for easy meals and have to be either relocated or destroyed, so I was glad my sighting was of a female and not the old man cougar. He had a warrant out for his removal. Lazy old hermit rancher that I am I would just as soon not have a lazy old cougar circling the place.

....

I love fresh fruit. In the mid and late summer I get down-right excessive when it comes to watermelons, plums, peaches, apricots, pears, cantalopes, and goodies like that. At our 3000 foot elevation and with our short season, we have not been successful with tree fruits. I must make pilgrimage to the markets for my 'need'. This summer I've noticed that, in our neck of the woods, there is little or no good natural or organic fruit for sale (seems the good stuff goes to the big city markets where demand is hot). And the fruit at the supermarket end seems to me to have hit absolute bottom in flavor, texture and freshness. Maybe it's just me, but this is the first year of my fifty years I can remember throwing away fruit after fruit because they were downright unpalatable, bland or plastic tasting, little or no juice, just a pithy nothing.

I remember fruit with a perfumed sweetness which would make my heart race, just like that cougar sighting. You could not eat such fruit casually. It stopped you, widened your eyes, made deep side recesses of your mouth cavity ache for some of the juice. Made you lick your fingers before you were done. It would catapult you to a special shaded spot surrounded by penetrating golden summer heat. Each bite would remind you of a kiss.

This fruit I remember was the best of life. And each perfect remembered one built on the remembered one before to make me into an insatiable hunter of fine fruit. The more crazed I became the more necessarily forgiving I seemed to become. Between those blissful moments with a perfect peach or handful of grapes I kept "wanting" to believe that the one in my hand right now was very nearly as good - but it never was. The true perfection was never almost, it swept you along like Arthur Rubenstein playing a Chopin etude. You went, never having to ask if you were being carried along, you went, swept, entranced. It was as if eating the perfect fruit you were yourself eaten by that fruit - at the very least the fruit captured you. Your memory of the fruit's taste gives it a life extending into future decisions, tomorrow's choices. If you seek another of its full flavor it has you, it owns you.

I am owned in part by every best and worst experience of my life. If I am driven to return to beautiful Assisi, Italy, if I hunger for the Eucalyptus and citrus odors of my California youth, if I think I hear the golden tones of black John my Minorca rooster of 25 years ago, if I dread certain terrible memories of my two dead sons, if an ancient fragrant memory of Rio

Piedras Guava and Carribean bread brings tears to my eyes, I am owned.

Through rhapsodic memories the best of fruit owns me. But it is getting harder to find. These days I know that many gourmet food catalogs offer expensive chances to experience heirloom (or old time) varieties of fresh fruit. For twenty to twenty-five dollars plus shipping you can get a dozen of such shipped to you but my experiences there have been less than rewarding.

.....

I once had two old grape vines - one was a Concord grape and the other was a Muscat. I want to believe that the compost I side-dressed with (along with the best wishes of native ghosts) accounted for as much of their occasionally fabulous flavors as did their varietal genetics. I learned from them that flavor reflects completely the growing season's character.

I once spent a few seasons with two grand old matronly apple trees. One was a King and the other a Gravenstein with a Spitzenburg graft. Their fruit was often to cry for.

As a child I played in orange groves and once stumbled onto a troika of small trees dead center of an extensive Valencia grove. Those three little trees produced a tangerine-like fruit soooo exquisite. It shames me to recall how many of those balls of nectarian perfection I filched and inhaled. I was caught, turned in and punished for the theft. But not before that fruit owned my ten year old soul and in vengence disappeared forever from my cognizant life when the orchard was bulldozed to make way for a subdivision. Had it been old Tibet I don't doubt that the orchard would have become a religious shrine to protect the mysterious power at the source of the flavor.

Do you think it's possible that I'm imagining all of this? That the fruit, or the memory of it, is an embellishment - a swelling trick of nostalgia? Will the cougar I saw become something like that?

....

Every occasion I had to drive past that spot where I saw the cougar I would slow and look up that hill - as if she'd be waiting to look back at me.

About ten days after the sighting and in very nearly the same spot I saw a strange looking animal bounce across the road in front of me. Being so keenly tuned to see the cougar again I perhaps had sharper more expectant eyes than usual.

It was a silver-blue cat-like creature about the same size as our German Shepherd dog. It had a bob-tail, longish pointed ears and its body hair was medium long. It looked like a large blue, long-haired, bob-cat. And it bounced as it ran.

I shook my head as I raced forward, hoping against hope to catch another glimpse of this apparition. I was rewarded. There on the hillside to my left standing between two large old sage bushes stood, in full profile, this magnificent Lynx cat looking right at me. I was enthralled. In a wink it was gone.

Then I caught myself laughing as I imagined a meeting of the cougar and the lynx.

They were sitting amidst the windfalls neath the 100 year old crab apple tree at the old homestead orchard above us. The cougar had an old wild turkey and the lynx chewed on a grouse and they spoke in mumbled sweet and sour murmurs as they ate - occasionally taking a bite of apple. The cougar said "I saw that crazy hermit rancher again last nite, let him look at me long enough this time. Figure I own him."

"What do you mean 'own him'?" asked the lynx with a mouth full of bird and apples.

"Oh, he'll be looking for me every time he goes by now. You know where, down by that old stone wall where we cross the road."

"Yeah, I know where you're talking about. But what do you mean 'own him'?"

The cougar just smiled with her nose as she ate the turkey and the apples.

So you see the lynx just had to come see for himself. And then I figure he came to understand. I may never catch a glimpse of either of them again but they'll both be watching me for a long time through the brushy cover.

....

Maybe it was three years ago, I'm not sure. But it was a Wednesday morning in spring because my brother Tony and I were heading together to the stock yards to try to buy a few heifers. I was driving the pickup and pulling our old gooseneck stock trailer. Must have been between 9 and 10 in the morning. As I rounded the last bend leaving the ranch we were both struck mouth-open dumb by what occurred. Neither of us would have or could have repeated the story had it not been for having each other as witnesses.

I have to tell it in a string, though it all happened simulta-neously.

He was huge. The biggest mule deer we'd seen. A perfect five point rack. He lept the barbed wire fence on our left and took two long gliding, hanging, air-cutting, ballet-perfect strides across in front of us and lept the fence on the right. That alone would've been visual treat and experience enough but it was only part of the picture. We both had our individual windows down and heard the skyward shreiks. I think that is what first slowed me and allowed we could get such a full and perfect view of the stag. And perhaps it prepared us just a little for the rest of the picture. To the left of the crossing buck, and hanging in the sky in a tangled breeding tumble, was a large pair of mature Bald eagles! And to the right, shreiking their territorial protests was a pair of Golden eagles. To triangulate this picture, imagine the great birds, each pair, at 50 feet ahead of us 50 feet off the ground and 150 feet apart and the buck ahead of us and center. The buck was gone in a time-freezing flash and we both looked at each other and whistled.

"Tony, If I had had a camera pointed in the right direction I could have had a photo of all 5 of those guys, they were that close together!"

"Yeah, bro, but you didn't and nobody will ever believe us."

But it didn't end there. Those four eagles, in pairs, continued to fly along side of us, the bald eagles on the left and the golden eagles on the right, shreiking at each other. We were driving at 25 mph down a winding road and they kept up with us turning when we did for nearly a mile and a half.

(As a side note: I did tell the story to a local wildlife biologist and he told me it was impossible because the eagles are territorial and goldens and balds won't share space.)

We saw it and the experience stays with us, owns us. And it happened along that exact same stretch of road where I saw the cougar and the lynx.

......

Remember the old adage about not seeing the forest for the trees? My point is that the old adage is incomplete. It tells us that we need to look at the forest as well as the trees. That doesn't go far enough. To experience the forest for all that it is you must know the flavor of the shadows certain communities of trees make at sunrise and how those shadows build slow moving lattice work frames to showcase penetrating fingers of morning sunlight in its warming and warning dance with the dew.

And you need to know that the reluctant lead dog regularly sneaks off (or limps off) to taste the evaporating dew while it still contains the sweet prisms of early morning light. To him it is like that Davis peach.

I can't explain it well but the wildlife sightings on our little ranch and the hungry memories of perfect fruit are for me related. They come from and go to the same spot deep inside of me. They are prelude and underpinning. And I am thankful for their passage through me. I am a fortunate man to be a part of where I live and to try to farm here.

Chapter Nine

And Dancing Cows

I must warn readers that the following story contains self-deprecatory and fatty acid content. There are no morals nor hidden truths. With this story I stand before you as a small man of questionable fiber (no surprise to many), a *Don Quixote de la Barnyard* (loved for it I'm sure), a *Hiram Holiday de los survivalists* (hide your beauties!) a man with a lullaby for a growl (who's he kidding?).

To give certain actions the questionable credence of rationale I need to offer this scary preambulatory story. In other words, this first:

Late January, late evening, no moon. My wife Kristi, daughter Scout and I sat in the living room area of the rough ranch cabin we call home. (In case you do not already know, we live on a remote high desert ranch with no nearby neighbors.)

Nadia, our German Sheperd watchdog, barks rapidly and nervously, Kristi swivels her head around listening. I watch. She's my ears for soft sounds as I have only one.

"Hear something?" I ask stupidly.

"I thought I heard a motor out front," she offered.

I got up and looked out the window. She did also. Nothing. Leastwise nothing to see or hear.

We both went into "Oh well" mode.

Maybe twenty minutes later Kristi repeated her alert.

"It's a motor, outside, just started up." She jumped up.

I smiled figuring she had a classic dose of "the wife's" proverbial "hearing things that go bump," and ignored her as she went to the window.

"Isn't that odd? There's a truck driving away with its lights off."

"Can you make it out?"

"No, it's too dark."

What with our two past experiences being burglared we shared a spooky moment.

The moment returned when we learned later that a remote neighbor, 6 miles away, was robbed that same night. It gave me a large dose of fear-engendered hairy toughness imagining someone looking in our window to see if anyone was home while Scout slept in 'Mommy's' lap.

Skip forward to Valentine's Day

Kristi and Scout were visiting Grandpa and Grandma Gilman so after the passing of flowers and kisses to and from all the ladies, being as I am my hired man, I went home to the ranch early to check on our calving cows and housesit. Late afternoon, everything fine, so I settle down at the kitchen table to write the editorial for this issue. (The one I'm not using, it's for the best - it was long-winded and full of poorly crafted insults directed at incompetent governmental bureaucracies, corporate greed, and the infertility of elected officials - all that and without even a half-hearted attempt at civility. Let's just pretend I never wrote it. The only reason I mention it at all is so that you may sense my pugalistic mood at that very moment.)

Time seems to fly by and before long it's eight p.m. The rain had stopped. I looked out the window; it was dark, no moon. I got myself a cup of hot chocolate and had sat back down to continue writing...

A noise.

A very strange noise, followed by mumbles.

It sounded like a low human growl, the sort of noise a haired-over leader might make in reluctant acknowledgement of the complete incompetence of accomplices.

And it was right outside my window. Under it.

A quick look and I saw them. The distinct silouettes of three large heavy-set men bent over to avoid detection and lumbering awkwardly alongside our dining room's lilac bush window!

I sprang to action. Or should I say shrunk to action? Dropping low I too lumbered to avoid detection and made it to turn off the light switches. The outside noises stopped right when the lights went off. Spooky confirmation. I reached up from my crouch position and got the high-powered flashlite from its peg and quietly - so quietly - snuck into the saddle room to get my rifle. I did not put a shell in the firing chamber for fear the clicking noise would alert the intruders and give away my position.

This was serious stuff. I should have been scared but my adrenalin pump had been primed by my imaginary editorial shadow-boxing with CEOs, bureaucrats, and realtors. I had a false sense of invinceability. I was EditorMan - faster than a Pentium chip, more powerful than a stock option, able to leap four strand fences from a well-placed nearby rock! That's what can happen when a middleaged man is left home alone in the dark with a fountain pen and an imagination.

I was ready. I jammed my feet into my logger slippers, opened the door and, bent over, scurried the few steps out to the picket fence. Once there I took a deep breath and stood up snapping on the flashlite and cradling the rifle.

I had loaded words on my tongue all in proper sequence, rehearsed in the warp speed domain of hero-mode, ready to spit them out as an order in some city-police-like staccato. But they stuck there and deflated when I discovered by flashlite that our cabin was surrounded - by cattle.

No burglars, no sneaky hulking hair-over criminal types. Just cattle. Lots of cattle. Our cattle. And they didn't belong in the yard. I had seen cattle walk by the window mumbling with mouths stuffed full of wintering ornamental shrubbery - not men. I felt like a complete idiot but my anger at the cows being out gave my energies some justification. This is bad. This is a mess. This could even be worse than surprising three burglars After all, 80 head of cows and calves pitted against one EditorMan gives the bovinus gaggle the slight edge in numbers. Three burglars to one EditorMan - the sneak theives wouldn't have stood a chance!

I remembered our little stable shed was open so I headed that way, flashlite making broad survey arches and seeming to switch on pairs of pink lights on the face of each watching cow. The lights were right where the eyes should have been. Then I heard it. It's no mistaking the sound of me muttering, grumbling, growling, forming verbal releases, wasting truly creative insults on my giddy fugitive cow herd.

At the shed we call our barn I flip on the light switch and discover ten head of fat cows burrowing their itchy winter heads into the primo loose hay I have socked away for my teams of work horses. I'm getting madder by the minute and walk into their midst slapping bovine butts and hollering. Two opposing cows try to kick me and get each other. They both beller and jump diving for the welcome safety of the dark wide outside. I stop to close the gate to the barn and my now muddy slippers stick to a clump of once beautiful hay drug to that spot on the floor by a cow unwilling to drop the stolen morsel until swept outside by her stampeding playmates. Backing up to shake free my slipper, the other foot is received, dead center, by an enormous celebratory cow plop. It was the size of a Toyoder steering wheel and the consistency of fresh cow plop. (Nothing else describes it!) I felt it close in around my foot and ankle.

I love horse manure and cow manure. They are two of the aromatic/tactile reasons I chose to be a farmer. But there are moments of circumstantial vermiculite when what we love becomes what we hate. I turn around angrier than before and almost get run over by a couple of heifers answering a "hey, girls, over here!" call from taunting lead cows in the dark.

I drop my flashlite in the mud, grip my rifle with mechanical certitude, and massage my forehead with the palm of the now free hand.

Isn't it fun how an aggravating inconvenient little catastrophe can quickly impose on us a sense of senseless urgency? I was on the brink of unjustified panic.

Palm massaging odd thoughts into my uncentered and discombooberated head I hear myself zero in on a desperation.

"Think man, you've got to get these cows gathered and you are on your own. Long ago you should have replaced beloved 'Sammy' the cow dog now resting in Border Collie heaven. Now is not a good time to be alone, you need the aid of a little command-hungry firey furball with that firecracker bark that cows move away from. But you ain't got it! Collect your thoughts! And for heaven's sake don't calm down! Now is the time for action not meditation."

So I walk with forced calm and a loose-fitting anger, back to the house where the rifle is stowed, boots are laced on, hat and coat are added and I go out to my pickup truck. I figured that I needed to make a wide circle of the farmstead and pasture environs to see what needed caring for first. Pointing the truck to see into corners where cows might be causing damage my suspicions are confirmed. They are everywhere and busy moving back and forth to everywhere. Over under the big Juniper there are actually two expectant mother cows fighting, head to head, over a few square feet of new grass. Unseemly. I'll get to them later. Then I hear Tom and Jerry bugling. They are the two Shorthorn bulls locked up over at the old corral. That's where I'll go first. I certainly don't want them getting out! I drive out the front gate and see a dozen cows on the county road.

"Oh great! Just great!" spills the sarcastic me.

So I drive cautiously as if to minimize my presence, (HA!)

and get to the corral before them but the girls decide to trot in front of the truck. I speed up, they speed up. Soon I'm doing 25 mph with cows, bucking and kicking and pointing tails towards the sky and clocking at least 30 mph. When they get to the common corral fence with Tomas and Jeraldo I park and walk past to open a gate figuring I will be able to gently herd them 20 feet to go inside the paddock adjoining the corral. Nope. Soon's they see me walk across the headlight's beam they bolt at a run - back the way they came from.

"Okay, okay" I mutter feeling somewhat like a naive voter whose winning candidate turned out to be without merit. The party-happy cows run right past the front gate and into the next gate which is government ground.

"That will hold them for a few minutes," I thought and went back into the farmyard.

I park the pickup with headlights shining on the hay stackyard and notice the bulk of the cows have busted the wooden gate and are lined up, heads in and eating, all the way around the three big haystacks. Each stack has 120 feet of circum-ference, measuring 30 by 30 feet at the base and standing almost 20 feet high, the stacks are positioned in an L shape within a square fenced stack yard. That quarter of the yard with no hay stack is home to the winter storage of the two JD buckrakes and the swinging hay stacker. Between the haystacks are eight foot wide alleys to allow a feed sled to pass. I'm struggling here to paint you a descriptive picture which might allow for you to "feel" what comes next. But it's too important so I include a little line sketch of the layout.

My brother and I and Kristi and Bud and Amy and Suzie and Harold and Kathy worked awfully hard on those beautiful haystacks. Though they will be fed out it still feels like a violation to have these surly cows force themselves on the silent sculptural beauties, eating holes in the perfectly groomed base of each. Imagine with me that you are gazing on three mounds which look much like chocolate-covered cherries from a candy box.

Only they are huge and thatched. Now imagine a wall of cow fannies pointing out from these shapes with tails wagging gently to match the chewing action of the burrowing front ends. It's not fair, it's obscene.

Confirming my entry into senility and the fringes of existential nonchalance I waded into that stackyard full of cows with flashlite waving in one hand and hat waving in the other whilst I yelled...

"What are you doing in here? You #@!*? cows! Get out of here! If you aren't out of here within five minutes I'll staple your tails to the ..... "

And they responded by moving a few feet away and continuing nasty disregard for my authority. So I picked up the nearest thing to me, which happened to be a flat bottomed shovel we keep in the yard to shovel snow, and slapped a couple of fannies yelling all the while. That seemed to work, they were moving away from me en mass but with some difficulty because in front of them everywhere you looked were more cows. So I yelled louder and pushed and punched with the shovel holding the flashlite pointed down or to the side so as not to blind them to my angry large presense right behind them! It was working! They were moving en masse!

What's this? They were parting and moving around a big red cow who had the audacity to stand in the middle of the lane between the stacks and look over her shoulder at me! She had the effrontery of a talk show host(ess). I tapped her on the butt and she took three quick steps forward and then kicked out sideways like cows will do. She kept moving but with an oddly familiar rythmn. One two three kick, one two three kick, first the right back leg, next the left back leg in perfect timing as if to ... Rhumba music!? I listened. I even looked up as if that would offer some clue. Nothing. Just one two three kick, one two three kick. Amazing. Oh well, at least they were moving. I hollered some more as we went between the stacks and turned right at the fence to go around the stack and down along the other fence past stack number three to the open gate. But it wasn't working because most of the cows were turning back into the center of the stackyard between stack number two and three! So I hollered louder and waved the shovel as if it mattered. The cows just wouldn't be moved any faster almost like they were up against some wall? Ah, and Carmen Miranda would not allow anyone or anything to break her rythmn. One two three kick, one two three kick.

"Oh so you want to dance do you?"

The unlucky cow with the fetching swing in her hips had become the focal point for my anger. I spied a stone reached down and picked it up and threw it at her. She bellowed and jumped forward and I felt something brush me from behind. I turned and aimed my flashlight and had only a fraction of a second to fall backwards flat up against the haystack! Cows stampeded past me. Where were they coming from? And then I saw her, no mistake, she was the only old heifer doing one two three kick, one two three kick. She went right by me and I realized I had been pushing the cows around the haystack! In fact they were still circling it after I stepped back. When I saw an opportunity I ran between cows and got to the outside of the procession. What was I going to do? They were still circling. I walked to the alley between the fence and stack number three where I had assumed they would have gone. Then I saw it. The reason the cows were turning back was because the truck head-lights hit them square in the eyes if they were to go through the gate. What an idiot!

I jumped the fence catching and tearing my pants and pulling a muscle. I limped, threatening myself all the way to the truck where I turned off the headlights.

I went back to the stackyard this time walking around and going through the open gate. I walked up the lane between the fence and number three and stood there looking, once again, at cow fannies as they gobbled on the sides of stack numbers one and two. I spoke with the force and tone of one who does not expect to be heard but who has always yelled from this fish crate in the park and will continue until judgement day.

"THIS, HERE, IS THE WAY OUT OF THIS YARD!
YOU WALK, like this," and I turned and walked, "RIGHT
DOWN THROUGH HERE AND OUT! IT'S NOT THAT
HARD" and I turned when I got to the gate and was
dumbstruck. Here came the cows. I quickly got out of the way. A
steady stream of cows, heading some where with determination.
Muttering, 'let me by you fatso, move it, what's your problem tub
of lard?' They were harder on each other than I was on them.
And then the stream stopped. Was that all of them? I walked back
into the stack yard to look around each hay mound. Yep, here
were two heifers, and there in the corner chewing her cud was
Carmen Miranda. She was looking at me.

"I don't need you," she seemed to say. "You go away. Can't
dance, don't understand women out shopping, don't understand
elemental geometry, can't differentiate cows walking by the
kitchen from Jack Nicholson look-alikes hunched over, why
should I listen to you. Go away, I like it here, I think I'll stay."
She had my number besides which as the rejected ridiculed male
my haired-over certitude was leaking out fast. So I went to chase
the heifers. They went easily. And right by me from behind came
Carmen Miranda, one two three kick, one two three kick. She
was right, I didn't understand women out shopping.

I followed them and repaired and shut the gates. While
working I could hear mooing and figured out that the cows in the
neighboring government pasture had called these stackyard dames
out. That's why they left when I showed them where to go. They
were worried that the calling cows had some better eats. The grass
might always be greener...

I finished putting cows away and found next morning that I
had done a pretty good job of it under the circumstances. But
that night walking back to the house and nursing my pulled
muscle I thought about what had occured. Kristi was on her way
home, what would I tell her. Hmmm.

How about...

...'you see honey, there were several truck loads of burglars
sneaking around the house and I knew I was outnumbered so I

snuck out and let the cows into the yard and stampeded them so they'd scare the meanies away and it worked but in the morning you're going to find some damage to the Lilac bushes..'

Naw.

When she got home I told her,

"Cows got out, took a bit of doing but I got them in."

Well, that's the story. I told you it wasn't pretty. And I told you there was no moral, leastways not the usual preachy sort I'm fast becoming famous, or infamous, for. The Monday after that weekend I went into the Journal office and the ladies asked me what I did for Valentine's Day and my birthday, which is the day before. So I found myself recounting the tale of the Valentine Rhumba. They thought it funny, in fact Suzie slid off her chair, repeatedly. Seems she was having a bad day and the story hit the spot, so to speak. I don't think the story is funny, it wasn't then and isn't now. But what do I know? So I thought maybe, if it has antidotal properties I ought to roll it into the aisle like a smoke bomb and see what jumps, or slides off a chair   ...or rhumbas.

Well. I got this wrote and then in comes a special letter from an old and dear friend who now lives in Bemidji, Minnesota. Back over a quarter of a century ago Marley Kaul and I shared a graduate painting studio at the University of Oregon. When you share a creative space and energy with someone, there is a unique sanctity and vulnerability there which may never be fully communicated to another. Marley knows a part of me. And I know a part of him. Marley and his wife Sandy had a baby daughter, Allison, during that time. Marley wrote to say that he had retired from teaching art and that Allison got married this last summer. Time travels. He shared with me a slide of an exquisite recent egg tempera painting entitled "The Horse Consoles Himself with Memories of Summer." He said in his letter, *"it includes your horse - sort of represents me - feet planted on the ground ... back to the wind ... white hair and looking back with fond memories!...thought you'd like to see you are in my thoughts."*

I thought, more than just little choked up, about Marley and Marley's words and then for some strange reason thought about my ring around the haystack experience with Carmen Miranda. You know that cow was in front of me and behind me at the very same time!

With his letter Marley is in front of me and behind me at the very same time!

It does not matter where we are in our lives, or in respect to our dreams, or as regards our age - each of us cross paths with sacred friends and dancing cows who, if we let them, will prove that all life is a ring around the mound.

Oh, I don't tell it so good. It belongs in a painting. For very long now you've all been so good to me, I hope you will forgive this one more piece of too much. Sweet dreams. LRM

Chapter Ten

Little Heroes

We are surrounded by little heros working like an army of elves towards, always towards. Good or bad, selfish or self-sacrificing, the work of our children shapes the future every bit as much as does our own efforts.

"I want to work with you Daddy. I want to help you feed the horses? Can I? Huh, Daddy, Can I?"

"Mommy! Daddy says I can help him! I'm going to feed Polly and everybody! Really Mommy I am!"

"Mommy will you help me? I need to make some 'maginary breakfast for my monkey. He's so hungry and he's too little to make breakfast."

"Why do you have to go to work Daddy? I want you to play with me. We can build a castle. That would be work. You can work with me Daddy. Okay?"

Raising children anywhere in the so-called developed world is treated far too often as a job, a chore. How tragic and destructive. What a loss of energy and opportunity. The direction our children's efforts take, from their earliest years, can be influenced beneficially by those adults graced with the opportunity to share early years with them.

Our children are little heroes because they face a sometimes monstrous controlling world where every day everything is more powerful than they and yet they persevere with wonder and energy and a boundless capacity for joy and grace. They are little heroes for their courage and strength. They are heroes because they so often save us from ourselves.

To allow ourselves to be inside a child's estate, a child's world, a child's imaginings as fellow traveler; this lends direction to their young work, their young efforts - and it can save us. Not a manipulative direction, not a *"no let's go this way"* direction. Rather a direction which comes from the growing strengths of shared wonders. Children will return to moments, thoughts, efforts which were amplified by a genuine shared wonder.

When any of us return, repeatedly, to a moment, thought, or effort which has been made attractive and holy to us because of shared wonder we build a saving tradition. Saving because, whether as a child or an adult, the return is affirming of life. A tradition because the returning path shapes who we are and what we care for.

Little people, both children and adults, need saving traditions. It is such traditions which preserve the memory of the saving deeds of God. Without that memory we are lost to ourselves.

Large people (those who make and see themselves as large - yes, me and some of you) need the humility that would make them (us) little. Otherwise saving traditions are invisible to us.

This last summer Kristi and Scout and I took a road trip to the city of Portland. It is normally a 3 hour drive. Approaching four years old, as Scout was then, Kristi was concerned about our daughter getting tired and cranky on such a long confinement in the pickup truck. Wide awake, looking at the moving scenery, and yabbering with Mom, Scout was the picture of contentment until her little brow furrowed and she asked,

"Daddy, are we here yet?"

We talked about her question being funny and worked to correct it.

Later, still driving, I thought about the irony. How truly accurate and important my little hero's question was. And is. I wondered, aren't we always here? And hopefully never yet? Where were we going? Here? Where are we going? There? or Here? or to 'Yet'? Surely yet is a place not as a subject but as an apology or a condition set to make 'there' less of a 'here'.

If you find the previous paragraph confusing you may disregard it completely. For you see the previous paragraph is a nearly perfect 'yet' and nearly worthless for that reason. Our subject is 'here' without the qualifications of anxiety ill-met or a watery apology.

"Daddy, are we here yet?"
"Yes, Honey we're here."

There is the parable of the good and bright young man who left his grandmother and their farm to seek his fortune in the city. The city not being interested in goodness, intelligence, and life-affirming thankfulness, the young man returned tired and disillusioned to the farm. There, of that first returned dusk, he rediscovered the 'homeness' of his place. On that evening porch his grandmother gently asked,

"What possesses good bright young men such as you to leave the beauty, tranquility, and security of a farm like this for the city?"

"Opportunity, I guess, grandmother," was his answer.

"Men would spend a lifetime in pursuit of opportunity so that they might one day be able to buy a farm such as this," came her reply.

There is a modern parable yet to be imagined where, instead, the young man answers the grandmother this way,

"I was sick of the sanctity and usefulness and beauty of THIS farm. Of its correctness and necessity. Of its tradition of saving ways. Of its self-ordained proximity to God. So I ran from those things to a morass of mankind where terrible beauty is found in curious juxtapositions and tradition bashing. Where life is raw and real and not a Norman Rockwell painting of perpetual quaint yawn. But that got old and I'm back to see if all of this farm, my youth, was like I remembered."

And the grandmother responds thusly,

"You were old and tired and stupid long before your time. When you were a small boy I used to marvel at your random sense of purpose and wild energy. And yet we always tried to control it and make you 'grow up'. We sure blew it. Now you're back here pining after this place and a life too long lost to you. Oh, it's not you, at this point that matters. Not any more. It's any children you may have of your own, or share space with. It's them who'll lack because of the brand of cynicism we burned into your spirit. Broken connections. Or maybe just connections never truly allowed. If only we could have enjoyed digging potatoes together."

"Daddy, are we here yet?"
"Yes, I think so, yes, definitely - we're here!"

This fall Kristi and Scout and I went into our little vegetable garden to dig a few carrots and potatoes for the kitchen. It wasn't for work. It was for food. And it was because our daughter wanted to. She was so excited to be walking to the garden with us to actually dig up some food! It was not a brand new experience for her. It was a return down a shared path of wonder.

"Look at this one, Dad! It's so funny. Mommy there's more carrots over here!"

You might think it's silly. But this moment and others like it are actually the center of the universe, the hope of mankind, the meaning of life, the garden of eden, the beginning. And the choice to "return" is central to the power of the moment.

Driving another time into town to be in the Christmas parade, my daughter Scout was beaming, I thought in anticipation for the parade.

"The parade's going to be fun isn't it sweetheart?" My lame 'adult' effort to share her moment with her.

"Uh Huh," she answered from distraction. Then she turned a life altering smile on me and observed, "the trees are watching us. Look, see they're watching us as we go by. On both sides. See, Daddy. They're watching you over there. They're watching me over here. They're smiling Daddy! The trees are smiling as we go by."

"Yep, they are baby. Just like me, smiling, and because of you."

"Yeah," she said slow, long and warm.

Goosebumps.

"Daddy, are we here yet?"
"Yes, Honey we're here."

May I offer this prayer?

God grant us the wisdom of youthful courage and the strength of silliness. Allow that we might cloak our wisdom in thankfulness and wear our silly strength like an edible badge.

Chapter eleven

Little Churches

"Complete possession is proved only by giving.
All you are unable to give
possesses you."
-- André Gide

I was asked to deliver a closing plenary at the
1999 Eco-Farm Conference at Asilomar, Califor-
nia. Below is both a distillation and an expansion
of those remarks. So the sum total of what you
might read here will in essence be the same if in
detail it differs.

Crisscrossing this park, this lovely setting for a confer
ence, I was struck by how many people I saw with
cell phones glued to their ears. Here at THE organic
farming conference it seems to say a great deal about our time.

I must interject that *we* use cell phones. They have helped us
because we can get no traditional phone service on our remote
ranch. So when I brought mine to this conference it was with a
measure of embarrassment because I suspected that the only other
person who would have one here would be the spy from
Monsanto corporation. I was wrong, there are hundreds of them
here.

We saw a peculiar and distinctive hi-tech posture everywhere we looked. The hand against the ear, head cocked to that side and free arm extended out gesturing to some waiting prospect as if on the stock market floor. I fear that the 'deal-maker' may become a trademark gesture for the new millennium. I would much rather see two hands cupped together as if in offering.

Perhaps these observations are why I had such an odd and appropriate dream last night. I have often been accused, and accurately so, of scoffing at those people who are panicked by the technology engendered form of Y2K fever. This dream came close to illustrating to me why people are so worried.

Imagine a great hall such as this, full of folks, and the opening bell of the new millennium has struck a tone. Immediately following comes a chorus of cell phone rings. Every cell phone in that hall buzzes simultaneously. And, in anticipation of that first monumental "deal" of the millennium, everyone anxiously answers his or her call only to find out that no one is there and the cell phones keep ringing and buzzing. As if choreographed by Busby Berkeley everyone then throws down their cell phone and stomps on it til the tones die. Now, with the collective realization of what they have done these saddened folk join hands and march, heads bowed, towards the cliff edge of oblivion as if so many Lemmings.

The lesson of the dream for me is that many of us are afraid of losing our toys, of losing our profit opportunities, of losing predictability. Many of us are unsure of whether or not we can cope if forced to be self-reliant.

I still see it as funny and somewhat silly.

Farming Reflections

I'm here to talk dirty - to lie to you - to flirt with you - insult you - and then leave.

I have been farming and gardening organically for 30 years. You should know that I have been doing this without certification or membership. I am not interested in the tight itchy under-

wear of compliance. I say yes to the rapture of passionate independent dedication. Dedication leads to community. Compliance
does not equate with devotion. Dedication may. Compliance
leads to corporate membership or fascist servitude.

I want to talk with you about why we farm. I'm not interested in political activism. Cultural activism is what we need.

Have we arrived? As my 4 year old daughter would ask, "are
we here yet?" Yes, right here. Whirling on the lunatic fringe
together. Look at us. Look around. New age funky doodle farmers, agrarian puppeteers, anxious tender fragile dreamers, pick
pockets, charlatans, deal makers, nervous scientists, frustrated
bureaucrats, corporate spies and media monkeys. All together.
And that is as it should be. We all belong here. But we're not here
for the same reason. I see two categories of reason for involvement. You can think of them in different ways either as;

profit versus hope,
or as power versus salvation,
or as money versus magic.

Hope, salvation and magic spell good farming, they spell
community, they spell self-sufficiency, they spell local self-reliance.

Great profit often spells sterility.

It ain't about profit with an "f" it's about prophet with a
"ph." We're talking about the beneficial acidity of hope. Why are
we here? Face it. We have arrived. The world is hungry for food
of good health, the produce of a land of good heart.

Translation to some: huge market potential. Huge. Shatters
all previous notions of food value. Starting tomorrow at noon
there are many billions of dollars to be made.

Translation to others: fantastic hope for the planet and
mankind. MAYBE.

I'm here to tell you that if you think you can have your
share of both great profit and salvation you are up the creek.

When was the last time you heard someone say "there's a lot of magic here and I want my share?"

We are in the midst of a wonderful Renaissance of individual human possibility but our culture is in denial. We hang on to the baseless precept that specialization is the guarantor of success. That success is financial wealth and nothing else. But success, true success, for the thinking, feeling, reaching human is a working spiritual wealth. We get that with diversity. Diversity is the guarantor, not only of success but of true success in combination with style, self-sufficiency, and self-awareness. Diversity is what we need in all aspects of our biological, social and cultural worlds. What we need are doctors who are farmers as well. Teachers who are farmers as well. Imagine; lawyer/farmer!

There are millions of hardworking people out there who dream of someday having enough money and <u>retiring</u> to a small farm to grow their own food and some to share. They are dreaming and working for hope. Aspiring to a life they see as salvation. Hungering for the magic of human scale work. Hungry for escape from compliance, from servitude to blood sucking consumerism. They'd "rather" be someplace else. Whether they know it or not they have replaced the modern bottom line ($$$$) with a new idea. Their new bottom line is a string of "rathers" - they are working not for money but for that which they'd "rather" be doing. There is a mountain of hope in this critical social change. It contains the seed of the best sort of cultural activism from which comes a winning future.

When we fight fire with fire, something gets sacrificed. When we fight the profiteers with their own logic and tools we lose. It is my contention that when we fight the profiteers with the current forms of political activism we lose.

Insidious biotechnologies, poison public food, a battered biosphere, a defacto criminalization of self-sufficiency, the ridicule of local self-reliance, the murder of genes, and a dying topsoil are critically serious disease symptoms which we currently attempt to treat with political activism. We should, rather, be working to rebuild our cultural immune systems and thereby remove any opportunity for these symptoms to exist.

With cultural activism our "rathers" guide us and the market for poison, destruction and consumptive membership evaporate.

Perhaps our insistence on political activism stems from our shortsightedness. So many of us behave as though the world started the day we showed up. That it started for us.

My good friend Jean Christophe Grossetete of France visited Yellowstone Park this summer. While on a designated tour he had a forest service guide who was a Blackfoot Indian. At one point he asked her about a certain landmark and she answered "my grandfather did not know and I do not know." I love that answer. It is four dimensional and it speaks to the foundation a culture gives.

Time is past for political activism. It is time for cultural activism.

Our goal should be VIBRANT four dimensional biological communities. And the road map to our goal is a buried treasure we might unearth if we could but understand our true motivations.

What we do is important. We are farmers.

How we do what we do is more important. Many of us are committed to true sustainable farming.

But **Why** we do what we do is MOST important of all. Contained within that <u>WHY</u> is an explosion of connected spiritual, economic and physical reasons which when woven together create community and afford us solutions to most of our great contemporary social ills.

And *'connections'* mean community. Communities with traditions rather than rationales. Biological communities connecting bugs to acids, connecting temperatures to human emotions, connecting songs to growth, connecting decay to fertility, connecting pollen to birth, connecting people to people, connecting all things living one to another - biological communities make up a fabric. Strength is in the tightness of the weave of those communities.

How tightly woven are we? Humor is a tightener. Absence of humor is often found in monocultures.

The traditional songs - are they still in us? Do we remember the growing songs? The magic songs? Can we be reminded? Will we be open to reminder and reconstitution?

The enemy of community is that which denies weave, denies connection. Political activism is reactionary and sometimes denies connection. (Sometimes this is necessary.) Cultural activism is constructive and preventative. It builds community - which is immediacy, intimacy, touch. True community is culture.

Corporate ethic and centralization (with the industrial scales they require) are the murderers of flavor, health, diversity, song, self worth and tradition. They deny weave, they deny connection, they deny community.

I believe it was 1987 (maybe '86) when I was visited on my coastal farm by a Japanese contingency headed by Tedeo Ichiraku, a founder of Japanese Organic Agriculture. He was in his eighties and came with a message. He had worked during WWII heading up the Japanese Agriculture Department and advocating industrialized farming. Seeing the error of his ways, he jumped that ship and worked on organic agriculture. Now in his twilight years he came to ask me to pass along a message to any who would listen. He said,

*"The problem is one of scale more than technology. The answer is small farms. Small farmers WILL farm organically. Large farms will **pretend** to farm organically if it is profitable, or if they must. Tell them small farms are the answer."*

Are we interested in big bucks or salvation?

He was right, it's not just a question of organics. It is a question of scale and "nature."

Do we want a farm that's a factory or a *church*? I took part in the Eco-farm bus tour and was enthralled by Tom Brose's *Live Earth Farm*. His random patchwork quilt of fifteen acres was home to great diversity and fertility and magic. It was a church.

And Tom's attitude reflected this in full rhapsodic measure. I will never forget the delicious embraced paradox of his full-body smile and soft-shouldered wonder when he said to us all "an acre of strawberries is a lot of strawberries." He went on to explain that his fifteen acres was all that he could see needing. It filled him to full with possibilities and salvation.

Our church is a wondrous manured, bug-infested, dirt encased, world of acids and decay and rebirth all in miraculous balance. Our church is our dream farm. Our church is fertility. The factory is sterility.

Our church is fertility.
The factory is sterility.

In 1998 of the 100 largest economies in the world the U.S. was first. Mitsubishi was 6th or 7th. And 51 of those 100 largest economies were corporations. They are the enemy because they value profit margins before life in any form.

In 1864 Abraham Lincoln wrote:

"I see in the near future a crisis that unnerves me and causes me to tremble for the safety of my country. As a result of the war, corporations have been enthroned and an era of corruption in high places will follow, and the money power of the country will endeavor to prolong its reign... until all the wealth is aggregated in a few hands and the Republic is destroyed. I feel at this time more anxiety for the safety of my country than ever before, even in the midst of war."

The way to successfully fight corporate ethic is through cultural activism. If together we come to "rather" not use their products they will disappear like the dragons of old.

Why do we farm?

Farming has always worked best as part craft, part mystery. While on the bus tour, I visited with Andrew Scott of Hidden Villa Farm, and he spoke of his happy observation that upright tool handles and posts had bird droppings on them. He wondered, smiling, what that really meant. The decision was made to add perches around the farm where birds would be encouraged to congregate. Perhaps Andrew is actively remembering and implementing a traditional song of farming?

On our ranch I use an open implement shed to stable my work horses. During the summer months, while I harness my teams of horses, I notice the tops of the stalls lined with barn swallows preening and watching. When I go to the field with a team of horses and an implement, those swallows follow in an undulating scattered cloud. It is bliss to do good work with a gentle contented team of horses on land of increasing fertility and crowned by a bug harvesting halo of happy swallows. As the morning sunlight shimmers off their flitting bird bodies and the pungent odors of soil and horses and forage draw me in, I know I'm where I'd rather be doing what I'd rather do. I know I'm in my church.

As Amigo Cantinano has been heard to say "you have to have places for the ladybugs to hide." You have to have those places because the ladybugs who help you need places to live. And you also need those places to complete the fabric of your biological community.

We need to develop direct markets in inclusive ways as full community as festival as theater and as new definitions of engaging convenience. We must bring everybody in and allow them to be who they are. We must stop constricting our community. And the marketplace should be the beginning. We must allow the politically disenfranchised common working class folk to stake a claim to their corners of this marketplace.

We need to understand how political activism and the defining and limiting semantics of conflict solution (i.e. sustainable, organic, environmental) serve to alienate and thereby constrict community.

Why do so many good honest organic farmers refuse certification and membership? This is an important symptom which we ignore at our peril. If we get together to solve a problem and one of us is driven off we've forfeited our opportunity for solution. Solution was in getting together.

Stop worrying about "organics" and the "enemies." Organics will more quickly rise to the forefront as the best logical and spiritual solution if we work on the construction and reverence of biological communities. Multidimensional communities. It is ironic the degree to which alternative farming politics comes to employ the tactics of big government and big business. With rules and regulations and testing and tolerance levels and 'transitional' zones and legal definitions. All to structure the 'least allowable', all to provide a road map for creative quasi-compliance, all to distance us from devotion and dedication, all to make of us factory workers, all to make of us liars. We can never beat corporate ethics at their own game. Never. We should be embracing, culturally, the rapture of passionate independent dedication. For it is in the independence, not in incorporation, that we paradoxically find community.

The time is past for political activism on agricultural issues. Political activism, on the alternative agriculture front, leads us to crippling alienation. It is time for cultural activism.

Central to any mapping of a cultural activism must be a solid understanding of the important differences between method and scale. We must not mistake "better methods" (read organics) as the solution. Organic agriculture, as a preferred method of choice for all of food production, can and should be a "result" - a wonderful "result." The premiere question is one of scale.

We want healthy food, healthy soils and healthy communities. We get these things by addressing the issue of scale. A land may produce food which measures as "clean" or organic and yet that land may be in poor health. Our goal should be a healthy fertile vibrant living soil. We do not have this with factory farming. We do not have this with monocultures. Our farms need to approximate gardens in their patchwork quilt diversity. Our farms need to be little churches. Common crude glorious little shrines to fertility and biodiversity.

In our political activism we ironically license and condone corporate organic factory farms. We play a dangerous game with legal definitions and forfeit the opportunity of the moment.

The agrarian lessons of Thomas Jefferson, E. F. Schumacher, Scott Nearing, Wendell Berry and others can, and perhaps should, be distilled to a simplistic charge; "we need a limitless supply of small independent diverse farmers. Therein lies our solution."

The opportunity of the moment? The world is hungry for food of good health. People across the board are worried about utility and food infrastructures. The hunger and concern is now, today, here. It is an economic reality. It's an emotional necessity. This hunger has created a void in the modern spirit.

The world is increasingly frightened by science in the service of corporate industry. People have grown suspicious of store-bought foods. Have we arrived? Yes, we're here now. We're here right now and we should, as concerned farmers and community leaders know what to do. Time is past for political action. Time is past to verify our individual and collective dreams. Our dreams won out.

A frightened hungry people in desperate need of clean healthy food, scientific accountability, and self-sufficiency skills. This is our opportunity. We can provide these things. And when we do, the change for the good, will be rapid.

It's time to get the stuff to the tables. It's time to eat. It's time to give thanks. It's time to invite the neighbor kids into the garden. It's time to get culturally active. It's time to make of our farms little churches. We've arrived. The water is beginning to clear. But we can lose this moment if we insist on continuing to squirt out the dark inks of political arrogance.

Community is paramount. Community, the result of cultural activism and complete inclusion, is the key to preserving and passing on seeds, traditions, ideas and even farms.

Community as a multilevel relay race. Overlapping generations and tightening the weave.

This is the dawn of the age of constructive indecency and political compost. The solution rests within the cultivation of "rathers." The solution is in our little churches.

Chapter Twelve

faith, magic and absentee ownership

The Minister's Horse, a Parable.

A wealthy greedy busy drinking man with vast holdings sought to address troublesome forebodings about spiritual insolvency.

He was looking for a quick, painless, inexpensive way to assuage his fears about the next life. It was brought to his attention, incidentally, that a local impoverished minister of good spirits and a powerful soul was forced by circumstance to make his rounds afoot.

Amongst the wealthy man's vast holdings was an expensive headstrong white stallion which no one could safely handle. The wealthy man did not personally know the stallion other than as its absentee owner. After a discussion with the manager of his livestock, the wealthy man instructed that the white stallion should be made a gift to the good minister. He wrote a note to that effect and instructed the livestock manager to attend the church service and place the note in the offering plate. The livestock manager was compelled to point out to the wealthy man that no one had been able to handle this stallion. Through his smile the wealthy man answered that the minister's success or failure with the stallion would not diminish the fact that he, the wealthy man, was making a substantial valuable offering.

When the good minister arrived at the farm estate to take possession of the offered stallion he was warned by the livestock manager about the dangers. The good minister watched the stallion from over the fence. To the amazement of the livestock manager, with a headstall and rope the good minister entered the pen. He stood quiet, eyes closed, praying and frightened but certain. The stallion snorted, screamed and charged around the pen for half an hour. When the stallion became quiet and approached the good minister, the headstall was put on and the two of them walked off, side by side, towards the parsonage.

The good minister was not a horseman. He was a timid man of faith and patience. And he was growing tired of the limitations of travelling always on foot. Over the next several days he became friends with the stallion until, fear in his throat, he decided to mount the stallion and attempt to ride. He first approached the horse and said, softly into his ear "I am not afraid of you but I am afraid of getting on your back as I do not know what I am doing. Thank you for your patience with me."

When he was on the stallion's back everything was fine and comfortable. Somewhat surprised he gently squeezed his legs around the horse's torso and whispered "Thank God." To this the stallion responded by starting, slowly, carefully, to walk ahead. The good minister smiled, leaned back and whispered "Amen." The stallion came to a sure stop. Over time they were to agree together that the command to go was "Thank God" and the command to stop was "Amen."

With the aid of his new stallion the good minister found his community of faith expanding. In the next town the tavern owner's wife was taken gravely ill and a call was put out for the good minister to come.

This same day and at the same tavern the wealthy man, having come by motorcar, had consumed great amounts of alcoholic beverage. He was sitting slumped in his car trying to nap when the white stallion came up at a trot with the good minister astride.

The tavern owner came out to greet the minister and was glad to tell him that his wife's fever had broken and that she would be most happy to see him.

"And what is this magnificent beast you ride?" Asked the tavern keeper. So the good minister told him the story while unbeknownst to him the drunken wealthy man overheard.

"...and so now when I say *Thank God* he goes ahead and when I say *Amen* he stops."

The tavern keeper was amazed and the two men entered the building to visit the ailing wife.

The drunken wealthy man fell to napping while he thought about the white stallion.

....

The wealthy man woke still dizzy from alcohol but anxious to be on his way. He knew he was in no shape to drive and he spied the white stallion standing quietly. 'Why not?' He thought, 'by rights he's still mine. If I'd known he'd be so easily trained I certainly wouldn't have given him away!'

So he climbed aboard the horse and muttered 'Thank God' to which the horse responded by walking off quietly. The drunken wealthy man was elated, giggling he said as a taunt to the horse, 'Thank God, Thank God' and the stallion broke into a gentle trot.

Leaning forward and burying his face in the horse's mane he shouted 'Thank God, Thank God, Thank God, Thank God' and the stallion broke into a fast gallop. The drunken wealthy man was euphoric. When he finally looked up he was startled to see that they were fast approaching a deadly cliff edge. "Whoa!" He shouted "Stop" he screamed until he finally remembered and said "Amen" to which the stallion slid to a stop within inches of the cliff's edge. The now sober wealthy man wiped the sweat from his brow and whispered "Thank God."

....

He was falling through the air to his certain death when he felt the good minister's hand gently waking him from his nightmare.

He was still seated in his motorcar. It had all been a dream.

The moral of the story is that there can never be absentee owner-
ship of faith and magic.

Within the parable of the minister's horse are several
representations.

The wealthy man can be seen to represent not only
absentee ownership but also large corporations and the corporate
ethic.

The foreboding of the wealthy man represents the recent
corporate concern for the twisting, turning, shifting tide of public
opinion.

The white stallion is a symbol of the natural world,
magnificent and difficult to tame and predict.

The good minister represents the best of the common
working class, poor people and especially small farmers. A man in
touch with his work.

The livestock manager can be seen as a representation of
the upper middle class, hesitant to care too much about what he
does because he is without ownership, without connection.

The joining of the minister and the stallion represents the
magical melding of craft to the natural world.

The dream state of the wealthy drunken man represents
the corporate fears of public opinion, the loss of profitability,
mergers, liquidations, downsizing.

The minister's awakening of the wealthy man is an
analogy to a future for hope.

As a painter I think visually and I react to where the
picture takes me. In writing I can only tell stories and 'talk' my
way to a point I want to make. I process information as I work.
Most of my best pictures and worthy ideas have come as personal
reactions to the process of painting or talking. And what I con-

sider to be my best pictures and ideas are too personal and seldom make it out to public viewing. The point is I NEED to be within the work, close to the work, in order to make it work - in order for it to be mine.

      I like to think about the example of the magic farmers. It is somewhat redundant to speak of them that way because magic is farming and farming is magic.

      Magic is farming not because of elements of illusion but because so much of the best farming requires a sublime faith in the unknowable. Because so much of good farming requires an up close knowledge of the workings of craft. You cannot make the card disappear unless you pay precise attention to the exact angle of your hand and wrist. The pleasure in witnessing magic is cradled in the wish to believe. The pleasure in creating magic is cradled in the knowledge that what they saw is grander than what you produced. You cannot count on the seed sprouting unless you pay precise attention to the exact condition of the soil. The pleasure in witnessing magic in farming is cradled in the wish to believe. To believe that the new born calf is part of a plan. The pleasure in a farming magic is cradled in the conviction that we are instruments in a grand plan which we may never fully understand. Yet even so our efforts are frequently rewarded tenfold. The farmer who, with an abiding reverence for the biological and spiritual mysteries of all life is able, time after time, to tickle the soil at the right moment and mix natural ingredients into the planetary epidermal stew resulting in increasing fertility and health-giving food - this with humility, patience, curiosity and intelligence, this is the magic farmer. Ask him to serve the soup.

Chapter Thirteen

Of Hermits and Harbingers

Counting beans that were,
beans that are,
and beans that are sure to be.

In blind pursuit of the "bottom line" we miss entirely the "top line." We work so hard to accumulate the means to buy 'stuff' that we frequently throw away or disregard our true assets. The truest asset we may 'own' is our self, that self which works to provide, works to create, tills, plants, gathers, stores, molds, appreciates, fixes, guides, parents, partners, loves, shapes, mixes, cooks, builds, displays, sells, and prays. Some would say we do not, cannot, own our 'self'. I must disagree. I believe we all struggle to take ownership of our self, in so far as ownership means awareness. It is a secret of success. And it begins with 'whole' accounting.

Farming ain't what it used to be. We heard that all the time, thirty years ago. Meant that the old general farm, with a family in residence joined by a mixture of livestock and crops, was a thing of the past. "Get big or get out" was the official war cry. And the government pushed hard to make it happen. So the large scale industrial farming model became the mainstream, became the orthodoxy. Now we hear it again, *farming ain't what it used to be.* But this time the lament (if that is what it is) takes aim at the poor struggling large scale industrial model. Now, instead of a war cry there are murmurs growing in collective hum, and those murmurs say "Downsize or get out."

Interesting ...

Our wealth is hidden within what we truly value. Size and scale and piles of beans can be critically important but not determinate of themselves. You can have the perfect size farm (for you) and a huge pile of beans and feel unsuccessful. You can have a frustrating distant dream of a farm and two beans and feel, because of your appreciative self, a tremendous success. If you 'feel' successful you are successful. If you feel unsuccessful....

What determines our success is what we value. I value my experiences and how they shape who I am. I value who I have known, and how I've known. I value the stories of my life. The big stories with grand plots, the complicated secret stories, and the little stories of vibrant lesson. When I give myself time to reflect I am showered by those remembered little stories and what they would tell me. Here's one:

Around the world with Marvin

We had a farm in the coastal range of Oregon, up in a high wet little valley tucked in behind a natural lake. A place outside of time. Neighborhood consisted of six dozen people on farms and homesteads twenty five miles from a rusty mildewed temporary vestige of attempted civilization called a logging and fishing town. The valley was a magnificent and merciless landscape within which this small scattered sprinkling of people proved that even here there was that third criteria beyond race and religion which human society, when allowed, would filter by. A criteria of animal insouciance, of neo-primitive survival. It comes of a people who have become intensely and perhaps even unnaturally natural. You may be of the hour's chosen religion and race and still be filtered out of society by your displayed attitude towards the trappings of class. Social accounting, where the intangibles frequently enter as debits.

Marvin Haskell was a jovial engaging disjointed hermit. Though by my description it may be hard to believe, Marvin was a real person. A crippled old wizened yet well fed curmudgeon who lived a mile up the road on his 300 acre timber/former dairy farm. In a tar-paper-covered single-wall fourteen by sixteen foot one room tool shed of a shack with no running water. All the outdoors was Marvin's outhouse, parlor, and workshop - the whole year long. His entire three-tiered rundown dilapidated farm environ was overgrown from the canary grass and weeds up into the wild Rhododendrons, Myrtle, Alders and beyond to the monstrous canopy of ancient Firs, Spruce and Cedars. Some of Marvin's trees were so large you couldn't see them unless you looked for them. Same was true of Marvin. No, he wasn't so large, it was that you had to look at him to see him. He blended into his background, his world. You had to watch for movement. Be looking right at him with a changing focus. And the movement you saw might be one of Marvin's two dozen dark wild Jersey cattle flitting around with the deer, bear and elk.

Marvin wasn't a chameleon or changeling. He was a stinky five foot eight whiskered red-pointy-hatted, double-torn flannel-shirted, cigarette-mooching, wood elf who, in his habitat, blended into his surroundings like an aboriginal hunter or a second-year carpenter's helper.

Marvin was a hermit but he wasn't antisocial. He was too good humored. He loved company. No, rather it was hermitage that was forced upon him because our society doesn't chance co-mingling with stinky, sticky, clinging 'inappropriateness.' Not twenty-five years ago, especially not today. And hobbling old Marvin, with two different decrepit untied shoes, because one ankle was so badly swollen, looked every bit like an urban alley tragedy.

Yet Marvin Haskell wasn't exactly down-and-out, he had clear title to 300 acres, half wildly fertile bottom land and half timbered hillsides. The timber was mostly oldgrowth Fir, nearly 5 million board feet of it. As a long since crippled-into-retirement logger Marvin knew the value of his trees. He was sitting on a fortune. As a wood elf and "natural" farmer he knew the value of his trees. He was sitting with a trust. Those two values were different, conflictive and Marvin always chose to preserve.

Because Marvin was crippled and alone most of the time, he turned to reading. He read everything and it all went together from thirty year old crime magazines, to Reader's Digest, to donated SFJs, to cheap science fiction novels, to great fiction. And when he needed some cigarettes or company, he'd hobble down the road to the first unlocked door. At that time I was conducting the Journal publishing business out of my farm house.

One summer day, by appointment, an older couple came to go through our reference library in pursuit of technical drawings of carriages. The man was doing buggy restoration work for the retired actress Kim Novak and needed some assistance. His pleasant wife was along for the drive.

He was at the dining room table with piles of books and she was sitting on the couch in the front room reading when the front door opened without a knock. I had grown accustomed to the osmotic comings and goings of several of my neighbors and that definitely included my hermit buddy Marvin.

When he would first enter a room Haskell could almost always elicit from unsuspecting new acquaintances a soup of three emotions; sympathy for his down and out state - disgust for his clinging mooching dirtiness - and humor for his twinkle and vitality. And for reasons beyond my understanding many women seemed to lean towards the sympathy and humor. It was a shared leaning for Marvin did love the ladies, especially if they showed him any attention. Before I could reach Marvin he had already taken his felt cap to his chest and with his dirty hand taken holt of the ladies paw, bending at the waist while he offered a long

drawn out self-introduction that was both apologetic and compli-
mentary. My natural reaction was to 'rescue' said lady as quickly
as possible, but alas I was too late. The bug of unreasonable
sympathetic attraction had bit her. I made it clear that Marvin
was not to pester the lady and she made it clear that he wasn't
pestering her. So I went back to my chair to continue reading
submitted manuscripts still within earshot of them for Marvin
was already seated next to 'herself,' his new 'lovely lady.' Her
husband perceived no threat and was lost in his buggy restoration
research.

I would look up whenever their conversation went into the
soft whispers and sink thankful when the volume rose enough for
me to follow at least the tenor of Marvin's talk. But this time
Marvin's theatrical nature and soaring imagination took to new
heights. He launched into a tale that began when he was 8 years
old and his mother sent him, in a row boat, down the Smith
River to its mouth at the ocean harbor at Gardiner to fetch a loaf
of bread from the store. (A distance I knew of nearly five miles on
the river.) When Marvin arrived he was attracted by the moorage
of a massive clipper ship, a freighter, loading for the Orient.
Curious boy that he was, he said, he walked up the gang plank to
have a look around. The boat was magnificent and he exhausted
himself looking into every nook and cranny. So exhausted was he
that he crawled into a lifeboat for a nap!

The soggy bent unlit cigarette stub dangled from Marvin's
lip as he stole glances to see if his quarry was hooked, and hooked
she was - mouth open, eyes wide, clutching his hand, she didn't
need to say a word. She was right there with that tired little boy
asleep in that lifeboat. It was difficult for me not to interrupt this
session if only by my chuckles.

Marvin continued his story: The ship set sail and the boy
was discovered. The men were all for throwing him overboard but
the kindly captain made him an offer. Work hard as cabin boy
and his life would be spared. The ship sailed to all the Pacific
Islands and to the Orient. All along the way there were fantastic
adventures reminiscent of *Moby Dick* and *Captains Courageous*

and *Kidnapped* and *Treasure Island* and on and on. Marvin managed to repeatedly save the captain and the ship. It took four years to finally reach India unload the cargo and head back for Gardiner, Oregon but not before the boy wonder was made first mate in a ceremony of laughter, tears and drinking during which the entire thankful crew showered him with appreciation.

The nice lady sat beaming for Marvin's accomplishment, she was as proud as a waiting mother, prouder than any girl friend had a right to be. Marvin was energized and told tales about the four year return which made Ulysses look like a supermarket box boy. When finally Marvin arrived back triumphantly to his home port, he asked his audience why she looked so concerned. He was prepared for her realization that it was all just a tall tale and struck momentarily dumb by her query;

"But your mother, what about her? She must of thought you were dead. Oh my that poor worried woman."

"No," said Marvin, "I just went over to the store and got the loaf of bread and rowed back home (five miles against the flooded swift river's current) and she was happy to see me."

Then Marvin got up and excused himself, said he had to go home harvest the squash and milk his cow. He was speechless for once. Didn't know how to handle the realization that this woman had believed every word he had said. In his mind perhaps he had failed somewhat to take full license of this creative opportunity. He could have been gone ten years instead of eight, fought three sharks instead of just one, and so on.

This story about Marvin telling his tale is all true. I treasure it along with several dozen other shared adventures. It is a piece of who I am. It is one of the accumulated assets of my life as a farmer and as a small time publisher. If I forget or refuse to count things such as this I dilute my net worth.

Truth in numbers?

Late winter 2001. If you want it or need it or are prone to feeling depressed there is ample current news to hold you down.

From disease to greed to incompetence to catastrophe. If on the other hand you want it or need it or are prone to feeling optimistic there is ample current news to hold you up. What are you looking for? That is one thing within your complete control, *what* you are looking for. (It goes to the core of how we value our self.) I am excited by the evolving social changes which allow and encourage that we look at things in new ways. Our outlook, how we measure things, and who & what we allow to measure us, these are intangible assets or liabilities. My focus, right now, is on assets. What do I value, how do I value it, what does it tell me about possibilities, what does it tell me about the future of my efforts? I am learning that we continue in the dark ages of measurement, of accounting. We are still counting just the beans in the bag with no measure taken of those which remain on the plant, of our ability to grow more beans, of the recipes we may apply to make the beans into soups and salads and necklaces. Who the beans remind us of.

As farmers OF choice, farmers BY choice, we have varying degrees of understanding about why we farm. It is because we are after a thing or things we value. Likely for many of us it is about a valued life and livelihood. We apologize to ourselves, frequently, that in order to achieve, maintain, and secure these things we value we must make a "profit." Within that apology, that less than certain stance, is a dangerous attitude towards social measure. It implies that we either value profit or we don't. And that one or the other position is not just a threat, it is down right evil. Many of us are zealots in our pursuit of profit for we believe it will make our dreams come true. Many of us are zealots in our disgust with profit for we believe that the pursuit of it destroys our dreams and diminishes that which we truly value. I have been frequently accused of being against 'profit' against 'wealth' against 'big' against 'success.' And less frequently as contradiction, I have been accused of being in favor of 'property rights' and 'the market place' and 'profit' and 'wealth' and 'success.' As if these positions, these measurements were all that mattered, or mattered at all. It comes from a linear way of thinking that ultimately abstracts what is actual so that we can connect assigned dots. It comes from measuring the 'bottom line' to the exclusion of any view of the 'top line.'

Sometime in the fifteenth century heavy thinkers came up with the basic framework for modern western accounting. A rule book for what is counted, what is subtracted, and what determines net worth. And accounting has not changed a fig since then. I suggest it is not because the system was/is perfect but because there was never collective vision and wisdom to question it - until now.

How about a new 'wholistic' or 'holistic' accounting system which welcomes the inclusion of all those intangibles that contribute to a true measure of a life? All those aspects of self and community that have everyday effect on 'what happens next'.

Making Things Happen

It was a couple of years later and Marvin Haskell had been spreading hints in obvious spots like poorly hidden Easter eggs. His sixty-fifth birthday was coming up and to him it seemed no one cared. It was my Kristi that spear-headed the idea of a surprise birthday party for the curmudgeon, "our" curmudgeon. Several ladies worked very hard to set things up. It was to happen at the school. Pot luck supper and cake and ice cream. A volley ball tournament. Horseshoes. A summer time community birthday party for Marvin. Nearly half the population of the little valley was set to come and that included a dozen or so folks genuinely fond of '*old dirty dirty*' as Marvin had been nicknamed. Actually the full nickname was '*Old Dirty Dirty the Ridgerunner*' because as a boy Marvin would run the coastal mountain ridges on perpetual errands for his demanding father. A peer of Marvins once told me '*Haskell* ain't changed a wit since he was 'bout fourteen 'ceptin' maybe for the whiskers. Looks the same, smells the same and even back then he was always moochin' cigarettes.'

To this day I don't know if Marvin was really surprised. He was pleased, I am sure of that. How could he not have been? Joanne had made a huge cake decorated up to be a Jersey cow and most everybody fussed over him and listened to his new bizarre story about how the Martians had come at night in their space ship, borrowed two of his best cows for breeding research, and returned them to him with an elegant multilingual thank-you presentation. And if we didn't believe him he could show us the two cows, he still had them locked in the barn and the Martians hadn't hurt them at all.

One of the patriarchs of the valley, Shorty Harrison, rose near the end of the party and gave Marvin a carton of cigarettes as a present. Then Shorty came to us, smiling, and said,

"Nice party. Marvin's either one clever bugger, or one lucky man. This is his second birthday this year!"

Marvin's no longer with us, but he is. He's an asset, so long as I let him be, so long as I hold on to his memory. So long as I count him into the measurement of this life. And now, to lesser and different measure, through the gift of these anecdotes, Marvin is one of your assets - if you let him be.

Sound of enthusiasm

Paradox. Today within corporate business circles new efforts are underway to redesign accounting and provide formula for including intangible assets into net worth. Ways to put a dollar value on copyrights and trademarks which reflect not only future earnings but also reputation, goodwill, and incubation values. We've been doing that right along with our farms, but without any sanction or permission. If we are smart and whole, when we measure our farming venture we take into consideration the intangible assets such as the quality of life it provides. Some are quick to note that you cannot affix a dollar value to quality of life. Yet we pay dollars for entertainment, exercise, vacations, diversions all with a view towards improving the quality of our lives. At that point it has a specific measurable value. If it costs us $2500 to be energized and/or relaxed and happy for a week's vacation, how do we measure a year of good living? I'd rather be out in my field working with my draft horse partners. I know of many people who would pay considerable sums to join me. I figure six months of good farming is worth $65,000 on the quality-of-life market. I would pay it, even after a lifetime of familiarity. Yet the old accounting would have my aggregate time spent working in the field to be less than worthless and that I should actually place a premium on finding ways for machines to do the work. I agree with my Amish friends who say the work is the reward. But something's happening. In spite of the old accounting which hammers at us that farming is a losing proposition, hundreds of thousands of new serious small farmers are joining our ranks each year. They are in smart pursuit of the lifestyle. In our own state of Oregon census figures indicate thousands of new small farmers and the data shows they are intensive operations with the majority involved in organic production for local distribution.

And it is just the beginning. We need millions more small farmers. And we'll have them because people are reaching for the intangible assets. The attraction and health of this life, and its product, is winning the battle for favorable public measure.

Chapter Fourteen

Source Work

O ver these last twenty-seven years we have seen, and heard of, many examples of boomerang lives. Specifically speaking tales of people who made passionate and reasoned choices to become independent farm operators only to leave after awhile to return to urban lives and then, with the passage of time and the morphing of rationale, to look longingly back at those left-behind farm lives. Their farming ventures were originally left for myriad reasons; failures with farming, pursuit of money, family, ideas, stimuli, fashion, health, romance, friends, therapy, etc. Often the diagnosis involved too much stress. What's most important in the story is that so many come to wish they had never left the farming adventure, or that they'd like to return somehow. There's information there about what manner of beast we all be. And it's important because we need to hold on to every single farmer while we attract and assist new ones. For these reasons we need to understand why we have these "boomerang" lives.

Industry and organized religion have whipped us into short-sighted followers. Industry says hurry up and religion says obey. We are told we must narrow our focus to a particular vocation or discipline and work hard on that one thing otherwise we will diminish our chances for success. The man who would be a cabinet maker and a botanist is asking for trouble. The woman who would be a dancer and a cellist is doomed. The child who would pursue automotive engineering and poetry is labelled a self-destructive curiosity. The best and most dynamic individual examples of diverse enterprise are frequently hidden from full view as an act of self preservation. We succumb to the pressures of family and community and deny the truth of our potential. We follow, short-sighted, the social edicts of our time. The result is often an unhappy life ironically even when we are immersed in the work of our choice.

Alexander Borodin was a Chemist and a Surgeon, but we perhaps know him best as a classical composer of the highest rank. We measure him today by the lasting power and grace of his music. By accounts he was a happy fulfilled man. Yet in his day, peers within the communities he worked would have doubtless wondered about his seriousness pointing to his various vocations as indication that he seemed spread thin. I prefer to think of him as spread thick.

I am a painter and a horseman and a farmer and an author and a publisher and an editor to name but a few aspects of my life. When I say all these things together in mixed company, no single aspect of my life is allowed full measure of serious regard. I am seen by some friends and neighbors and strangers as diluted and mixed up, someone who hasn't been able to make up his mind. I offer the example to illustrate the point. In my case, my life is so full and busy and dynamic that how others might see the larger shape of my efforts cannot and will not change my course. But for others struggling to design a life and make vocational choices, peer or social pressure often wins out.

So, for the beginner, the turn to farming frequently takes on the demanding aspects of a discipline requiring absolute and solitary devotion if for no other reason than the assurance that this is the only way to make it pay. Push, push, push.

"We're market gardeners and nothing more but certainly nothing less. We need more broccoli and carrots and fewer weeds and it must happen TODAY! Get everything else out of the way. Time's running out. There are bills to pay. Why can't we get paid what this stuff is worth? Why is it so hard to get good help? WHY, WHY, WHY?" This is unfortunate and deadly.

We would like to make an argument for diversity of human pursuit. And lay the claim that such diversity would protect the valuable resource we know as our growing farm community. Protect it by allowing for a human scale to our endeavors, a human speed to our endeavors.

If you are planning for, or well into a life of farming, don't shut out other interests, other needs. Find ways that the farming may bridge and invite and allow time for other pursuits connected or not. Play the cello and pick the beans, be the pastor of the church, a small engine mechanic and raise sheep, write for the local newspaper, make knives, be a basketball coach, and raise grapes. To do less than allow your interests to flower is to literally stunt your growth and perhaps inevitably to regret the farming.

With agriculture, to always push for maximum production and profitability with one or two crops and a specialized target; that's industrial by definition.

To gracefully gather, nurture, bunch, pile, trim, nap, digress, nudge, cover, choose, hum, stir, read, follow, measure, wash, allow, pick, and water, that's farming by definition.

Industrial forces frequently deaden the human spirit.

The forces of a true farming should allow the human being as source, as work, as complete. But the farming can become no less stressful, no less hurtful if our attitude is intense speed, the false efficiencies of scale, and assembly line thinking. We need to give our farming room to breathe and fulfill itself. We need to slow down and spread ourselves out as thick as our interests might allow. If we live or die by our milk check or our raspberry income or our U-pick income or by our CSA receipts, we are begging for stress and misery. If our milk check is offset by a few dollars earned working Saturdays at the local bookstore and by the sale of a quilt through the mail and by a part-time nursing gig, life will be more akin to a hand woven blanket than a plastic bag.

More than a dozen years ago I visited the late Parker Sanborn in Maine. He took me around his lovely Jersey dairy farm and explained that life had become better for them when they went to seasonal grass dairying, taking winter's off, and throwing away their ingrained notions of high production cows opting instead for long lived cows and better quality milk produced with less labor and less cost. His careful observations and thoughtful inquiries had revealed that so much of the efforts went to pay the costs of so much of their efforts. And then they realized they didn't need to produce their own hay, they could trade a neighbor for hay, take one more step away from stress and put time into spreading manure on the pastures with the Percheron mares. This gave the family time for other interests and the harmony and regenerative aspects of a farming life. Parker had found the recipe for true success and it required all natural ingredients blended or folded slowly with a long wooden spoon.

My take on Parker's recipe? Let the farming portion of your life's adventure be good to you.

If the whole of agriculture is like a watershed,
its rain and melt-off
creeping through hungry drinking roots
and the escaping moisture
forming streams, creeks and lakes
which shape the landscape
as they ultimately rush to join one another
and form powerful rivers
racing to merge

with the huge expanse
of the oceans and seas,
the work we might do as small independent farmers is like the rain,
like the melt-off,
like the unadulterated streams and creeks and the pristine lakes.
The work we might do as small independent farmers is source work,
it is about beginnings,
corny as it may sound it's about creation and adventure and life.
It is at the core of all the best which is and might be for the human
     condition
and the planet health.

Most modern men and women of the developed world are far removed from source work. They are along for the ride that a closeted life might provide. They often die tragically without having been involved in the shepherding of life force. They live insane lives in tightly congested insane communities. And they *wonder* after their own health, their own sanity, the education and development of their children, and the degradation of the natural environment around them. They wonder after these things but they put the larger urgency and negative energy, the worry, on a list headed up with income, possessions, appearances, associations, and position. They believe that if a life at the source were available, it would have a very big price tag. To own a farm, and work as a farmer requires, they erroneously believe, a great deal of money. To get that large sum of money, they erroneously believe, they must continue to live and work in the tightly congested corporate insanity, the same insanity which does not want to let them leave. They allow themselves to be trapped in a circle.

Assurance is a nebulous, shapeless, ever changing often negative thing. We want assurances that everything will be as we believe it should be. Or, that failing, we at least want assurances that if we work THIS hard we will end up with THIS as our reward. Someone or something offers us up the idea of assurance, "yes, senor, I assure you it will turn out this way." Churches, bosses, schools, doctors, friends?, pop idols, pimps, pushers, dog trainers, dance instructors, lecturers, teachers, writers, editors, mothers, fathers, and realtors all want us to know just exactly how it is and will be; they offer us good or bad assurances. We proceed and *'lo and behold'* it doesn't happen as it should have! What happened to that assurance? It didn't evaporate. It never existed as a thing, as a guarantee or warranty or certainty to begin with. It was a promise half made. Yet, men and women still seek and are governed by offered external assurance. *"I just want you to tell me it will be okay."*

If it is an understood goal that an individual is to become a farmer/craftsman, he or she must first be an individual, independent, self-*assured*. That is not to say or suggest self-deluded. That is to say someone who believes, realistically or otherwise, that what needs doing can be done and is worth attempting.

And someone who does not need constant external reassurance.

As farmers and craftsmen we work at the source and on the source. And we are the source. We are artesian if we allow ourselves to be. We bubble up as if out of the ground, as if from some mysterious hidden pressurized fountain head. We are at the beginning of the flow. And at the same time we cup the flow in our hands and drink from it.

And we feed fertility and we are fertility. We hasten reproduction in plants and animals. We, ourselves, reproduce. We set the stage for multiplication, division, addition. We use subtraction as a sculptor would use a carving knife, to shape and design and suggest.

One foot flat to the ground, all weight on the opposing knee, leaning forward and over the anxious plant, palm of the left hand caressing the spikey curvature of too adventurous a spring growth, right hand clipping and shaping, a picture of craft. That brain is filled with an experience-driven understanding of the likely result of such effort. The air is filled with warmth and small sounds and odors which think they are fog-like. At the more appropriate time, this farmer knows he will return to the plant and clip long flowered stems of the herb to gather into a bunch for drying. Cause and effect, purpose and outcome, resolve and result. But sometimes, perhaps even often, not so direct, not so easy, not so comforting.

One day the egg production falls off by half. A few days later a hen dies and several are looking ill. And on that day the old stallion walks with a staggering difficulty as if his spinal cord is pinched. These things are tended to with diligence yet with little gain. Add to this the broken drive chain, the failing hydraulic pump, the irrigation pump bearing's new scream, the mold in the grain, and the long gash across the palm of the hand from the errant barb wire.

But the clover and the grasses are coming on very nicely and the new hives will be set soon. And all the materials are at hand for the work to commence on the new pig house. And the fence rail is flattened on each end and drilled to received the spikes which will hold it to the waiting fence posts. Working alone this farmer balances the rail on two temporary nails and backs up to visually check for level. Looks good. Feels good. Nail it.

Yet the dry bean crop leaves curl around the beetle chew marks and begin to wither. And two of the turkeys are found dead with obvious signs of hawk or owl. One of the work mares walks with difficulty on what is discovered to be an abscess. And the well comes up dry...

The reserves from the sale of steers and herbs come in handy to drill a new and better well. The potato crop is outdoing itself this year. Glorious is an appropriate word. The lambs are gaining quickly on the legumes and the new four point barb wire, stretched tight at ground level, seems to have slowed the neighbor's dog. A replacement transmission for the old hay truck has been given by a neighbor. And the old favorite work mare has an easy birthing giving what will surely be her last foal. The farmer's daughter names the perfect new foal 'Pecos Bill'. The foal is a filly.

The farmer instructs his children that they must never tell

anyone that they are saving back seed from their own plants. They ask why and he tells them they will know the answer another day, it is enough now for them to understand that they keep this secret so that they may keep their farm.

A neighbor becomes distraught. He has tried with unabated passionate attraction but, after two years, it just seems like things will never come to profit, to resolve, to fruition, to calm, to competency. He knows what he wants. What he must have. And it seems that the only thing preventing his arrival is his own inadequacy. He wants his farm to work. And he wants to feel safe and competent with his team of work horses. He's read all the books, attended clinics and workshops, and still he has a constant struggle each and every time he goes to harness the team and go to the field. He's exhausted by a combination of anxiety and fear, frustration and confusion. It doesn't come from the knowledge base. After two years of harnessing and hitching, and many hours of actual field work he has a good grasp of the mechanics. He knows how to properly adjust the harness, how to fit the collar, how tight to fasten the hame straps, the proper angle for the traces, he knows all these things through trial and error and correction and repetition. These details are not his problem any longer. His problem is that he's unable to relax with the horses. He knows from too many bad experiences that they will try to run away any chance they get. So he's constantly wary, constantly reading their every move, twitch, glance. And they, in turn, are doing the same with him. And when he tires, his wariness turns to anger, anger directed towards the horses. And when they tire, their spookiness amps up and is peppered with sullenness.

So when his cabbages begin to show worm damage, these *horse fevers* amplify his anxiety about this turn of events. And when he strips the threads on a unique square headed bolt which holds the knife head guard to his horse mower and finds out that this bolt must be manufactured one at a time in a $65 per hour machine shop he figures he's had enough and goes to see his helpful neighbor to tell him that he's beat, done, washed up, through with the adventure.

And the helpful neighbor says,

"Not yet, you aren't. Here's what we're gonna do. While we hold everything still, you are going to turn 45 degrees to the left and come out a new man. You have cultivating and mowing to do, right? I'm gonna show you right now how to make a service-able square head mower bolt out of a round head plow bolt. Then we're gonna harness my old girls, Bess and Frog, and you're gonna drive us on my wagon over to your place. Before that though we'll load up four feeder pigs which you'll pay me for later. They're gonna get all that produce you say ain't gonna make it. Once to your place I'm gonna harness up your Tubby and Rocky, hitch them to my wagon, and come back here. I need more speed and you need more assurance, so we'll swap horses for a couple of weeks."

The borrowed horses put the young farmer at his ease. And the problem horses in the old farmers hands weren't a problem because he expected them to behave and he let them alone, all at the same time.

At the end of those two weeks the old farmer said,

"Hang on to the mares for a week or so more."

And the young farmer was grateful because he was loving his work and his feeling of competency and calm.

When it came time to swap back horses, the old farmer said,

"I think you should consider picking up a couple of wean-lings you can start. Play with them, hang harness on them, ground drive them around. When they get around a year old, hook them to something light and start driving them. The two years it'll take for you to get them ready for farm work will also finish your own education. You know what to do, you just have to lighten up and do it. These geldings may always be a little jumpy with you, and they'll always make you itchy. Pass them on to someone else while they still have value and get yourself restarted with young horses."

From even just a short distance away, it would appear that nothing had changed for the young farmer. It would also appear that everything had changed and that this was no longer the same man. Days progressed as before but with a lighter step, with smiles, with occasional whistled tunes, with fattening pigs, with comic young horses. Not much had changed in those ledger columns where income was recorded. But the young farmer had become artesian, he had become source. And the older farmer had the opportunity to feel his strength with the faster team and to realize more money for the feeder pigs than he would have got from the market. That and he had kept a good neighbor, one of great value to him in the future.

Though USDA analysts and theorists will say otherwise, agriculture at its core is not changing and the people in agriculture are not changing. It is that federal institutions and agri-industry have for decades wanted agriculture to be something nearly opposite of what it always has been. They have wanted and needed for it to be industrial process fit within the confines of corporate accounting.

Today dramatic new successes with small acreages conceived and managed by aggressive, bold and obviously intelligent INDIVIDUALS, has the purveyors of the corporate industrial model scratching their heads and running for political cover.

Agriculture is, by its biologically incurred definition, a craft-based stewardship of chosen segments of a naturally reproducing world. Agriculture is people gathering and planting seeds so that they might best reproduce and multiply allowing a gain in their number which may be used for seed, food and fiber. Agriculture is about livestock husbandry. Agriculture is about fertility. Agriculture is about source. People at the source, people working. And, if all is as it should be, people enthralled and exhilirated and sustained and rejuvenated and rested and assured in that source work.

Chapter Fifteen

Workhorse Diary

From any distance, our world has become a confus ing dry mosaic, a deadly patternless swirl, one with no adhesive and in which the pieces don't work together well. War all over the planet, terrible new pandemic diseases, widening gaps between the rich and the poor, widespread hunger, a frightening pattern of ever new and more destructive natural disasters, the dissolution of the arts, the industrialization/corporatization of agriculture and health and science, the repeated convoluted public sale of justice at the highest and lowest levels, corruption and stupidity at every level of government, and terrorism. What's a person to do?

In my case, within this small insignificant little life that I have been lent, I would go stark raving nuts if it weren't for the assurance, vitality, challenges and poetry I feel at my finger tips. I turn to the working spheres I have chosen, my family, my farming, my art, my writings. And within each of those there are disciplines or aspects which invite my continued involvement and allow me a chance for place and growth. Within family I currently feel the transitions from father, to curiosity, to friend, to grandfather. Within my art and writings I feel expansive, generous and collective all at once. And within my farming I feel charged with new ideas and hopes. In farming I am blessed to have disciplines, such as the working of my horses, which return me time and time again to patterns of powerfully effective though small membership in a difficult to define little working system, no, better to say working way or working world. It gives me a feeling of insulation and connection. It makes of me an independent individual and a dependent individualist. It gives me an organic citizenship in the whole of the wide world without compromise to man and with allegiance only to goodness, fertility, splendid humors and shared abundance.

If the world is badly messed up these days, don't we need detailed internal views of little working disciplines which connect us to a reality we would want to sink our teeth into? I believe so. Here's a taste of one of my little realities.

Working horses day after day the routine wants to remain constant but it varies, and it can be different and yet similar in important ways. Here's a narrative as personal diary with added observations.

The sun's not yet up. The sky is a cold silver-tinged with reds. Even before I step outside, I can see from the kitchen window that all eight horses have their eyes glued to the door of the house in anticipation. They are looking for me, for my approach, for my errands of service to them. As I walk from the house, towards the shed which serves as our barn, the horses begin to nicker from their night pens. I slip from a low down deep throated hum to a soft whistle and back to a hum again, this morning it's an Argentinian Tango. 'Lucky', the Australian shepherd, follows me, excited for responsibility. The horses bob their noses and shuffle front feet as if to say *'it's about time'.*

Inside the barn/shed I pitchfork large quantities of the loose hay into each double manger. The two-foot deep mangers run across the front of each double tie stall which measures ten feet both directions. Each double tie stall is partitioned from the next by a planked half wall. The floor is packed sand. For fifteen years I have promised myself a real barn with a wood floor and well placed harness storage crowned by a large hay loft. But finances and circumstances haven't allowed such an improvement so the horses and I make do. No, excuse me, it's better than that. We are thankful because what some might see as a crude open equipment shed temporarily converted to stabling for work horses we have enjoyed as an airy and well lit little barn. It has served us economically and well. I lean on the pitchfork and look around wondering how I will remember this building...

Horse's nicker louder and I wake up to the job at hand. First I go outside with halters and lead ropes to the pen which holds the two geldings. They are in a rail enclosure just 100 feet from the barn. It's their night accommodation. In the pen each gelding stands close by, nose forward as if aiming at a coming halter. I smile thinking about how these daily routines work so well to build the best working partnerships with the horses. Halters on and off everyday. Horses become not only accustomed but, by association, actually pleased with the process. When it comes to the haltering they each know that it is the first step in going to their stall and eating, so they lower their heads and reach their noses forward for the positively associated halter. And when day is done and halter is to come off, once again they lower their heads and now rack or rotate sideways, ears over towards me and down, to make it easier for the crown strap to come off, anticipating the comfort. They are happy going in to the barn and they are happy going out. An excellent indicator of a system which works.

Both horses haltered and with lead ropes in hand, we exit the pen. They are anxious but they know, because I never allow exception to the rule, that they must walk behind me and on a loose lead. No bolting, no dragging back, no stepping on me. We walk together easy and fluid. It's either that or I get cranky and dream up clever ways of making their rule violations uncomfortable for them. They elected me leader. It may have been a rigged election, but I am the leader nonetheless. If I fall out of favor and dominion over them, it is my own fault and I must work to regain the leadership position. Otherwise, it is my contention, there can be no safe working relationship.

So we walk calmly to the barn where I snap each one to their individual secured manger chain which is bolted beside the 2 x 4 grain box. (These boxes are anchored in at the left and right sides - or opposite sides - of the manger, one each for each horse.) There is enough slack in each manger chain so that the horses might reach the middle of the manger and not much further.

The chains have big heavy bull snaps. I don't want them to break. It's not because I'm afraid my horses will try to break them but because accidents may happen and I prefer not to have loose horses in the barn. I occasionally use ropes instead of chain and sometimes *panic snaps* which allow that a twist or quick down-ward jerking motion will release a tangled endangered animal. These I usually relegate to trainees or new horses I am unfamiliar with. Just a small dose of caution which may one day pay off in a big dividend by saving a horse's life. But to date, after 30 plus years, I have not had to use one in an emergency.

Geldings are snapped in and eating the morning hay. I flip off the electric fence charger and head back out to fetch a team of mares. With the girls you never know how the day's chemistry might stack up, one day mellow, the next day sullen or cranky. Today, with the first two, it seems they are mellow.

There are two pens, wired with electric fence tape. Because I have four teams of horses in for haying, and because one of the mares beats up any horse she has to share space with, I needed additional enclosures and have set up a temporary electric pen arrangement for harmony and my own convenience.

This season, in total, I am working five of my mares and three geldings. Four of the mares work in teams or as three or four. The fifth mare has issues, not with me or any human but with other horses. Before I bought her, she was made a pet of and spoiled terribly. What she wanted she got. She works well with her gelding teammate as long as I am around to warn her off her worst tricks. She listens to me. I make sure of that. I've had her a short time and am determined to trick her into changing her attitude and nature. Right now she gets a pen to herself and occasionally is put into a large box stall for isolation. What I'd like to find is that she develops an attraction for one of my other horses and wants to be with it at all times. When this happens, I will use it to advantage.

We are only in our second week of concentrated work. Not until each horse is putting in a full 8 hour work day will all the minor anxieties level off. But one thing shouldn't change, each morning every horse should be anxious about getting to the feed. If one shows no interest, this could be an important first sign that something is not right. It might be protracted fatigue, it might be sore muscles, it might be any number of physical ailments. Best time to deal with potential health problems is at the first indication.

But this morning everyone is alert and salivating in anticipation of the clover/grass mix hay and the grain they expect is coming. I intentionally choose the pen with the two younger mares because I want to avoid yesterday's little incident.

I had gone after the older mares first, haltering and opening their wire pen gate and leading them towards the stabling when, to my surprise and disappointment, the younger ones pushed through their wire enclosure. It was as if nervously watching me open the other pen they figured out that the electricity was off. I shook that thought off as ridiculous, how could any horse come to such a conclusion? The thought nagged at me though, so later yesterday I tried a series of little experiments and determined conclusively that one of the two younger mares would watch and if she saw me handle the electric fence wire tape in any way she would then lean hesitantly against the wire herself. If she felt no shock she'd lean into it until it snapped! I have trained this young mare from a yearling and have always marvelled at her intelligence, so my surprise was somewhat muted. I made the decision to outsmart her by always either leaving the charger on or removing her first and putting her back in last.

Both haltered, I toss the wire gate out of the way and head through the opening. One of mares pushed forward to get past me and I gently swat her chest with the loose end of the rope, she backs up. The smart mare makes her move, head down, for an alfalfa shoot at the lane's edge. She catches me off guard and I get drug half a foot. "What are you doing?!" I grunt as I jerk her lead rope and finally regain the composure I like to brag about. When we get to the big doorway of the open barn they, of course, tense up in anticipation and I use the moment to remind them of our election results. "I'm the leader; you follow, remember!" Turning to face them, while switching leads in my hands, I stop them at the door.

They can now see the geldings happily munching away in their stalls. Facing them down and using quick matter-of-fact little jerks on the lead ropes I repeat the "Whoa" command until, when they come to fully recall the routine, they relax completely to accept that we aren't going inside til I say so. When I see the heads drop slightly and the ears go limp I lead them in. This little exercise, I have found, whether on the lead or when ground driving back into the barn after work, pays huge dividends to reinforce leadership and frequently results in horses who will stand quietly for me in **any** circumstance.

They in their places in the stall, I go back outside to get two more mares. These two remain wary of the fence, on or off. And they've spent fifteen years with me so the routines are well understood. We make a quiet uneventful little walk to the stalls except for a wave of emotion I feel to think that one day these two very fine horses will, as others have, be too old to work with me. I will miss them and the assurance they always give me.

Last I retrieve the separated mare and gelding, this morning tieing them side by side in a double tie stall. I will be working in the barn and know that a repeated word of warning will forestall the angry mare from pestering her teammate. If I were to be leaving the barn for an extended period, I would choose to tie them in separate stalls.

Up to this point none of the horses have received any grain.

This is intentional. I want them to have chewed forage in their digestive tract before they start to eat their grain. In this way the grain is slowed in its passage through the equine stomach allowing for better digestion. Long ago a visiting veterinarian pointed out the whole oats apparent in the stabled horse manure. He suggested this timing routine (hay before grain) and a preference for rolled grains over whole. I have followed this suggestion for over thirty years with good results.

So this morning I go to the locked grain room and remove the lid from the galvanized garbage can I use to store the COB (corn, oats and barley with molasses). At the first slight sound of the lid lifting a chorus of pleading nickers are aimed at me. I never tire of the sound, that deep soft edged percussive staccato hum, simultaneous from several horses. It comes as a full acknowledgment of my place in the relationship. They seem to say "Yes, you, please. I'm ready, can I have my grain now, first, before the others? I'm over here, where do you want me to stand? Please don't forget me. I need you to remember me, I need you to bring that grain to me. Bring the grain now and we'll see how the rest of the working day shapes up."

I fill a bucket with the grain and go from stall to stall, grain box to grain box, measuring out amounts I deem appropriate for each individual horse. Inside each wooden grain box there is a small salt block. I slide it to center of the box and pour the grain over the top of it. I have this unverifiable suspicion that this simple act slows the horses from quickly anxiously scarfing down the grain, spilling some on the stall floor. Watching their nose, lips, and tongue navigate the grain piled around the salt block, it seems like the rhythm and speed of their eating are natural.

Then I pause to listen to them, munching, snorting, breathing, pushing the salt blocks around the box. The view of their lovely forms, comfortable in their stalls and with their routines, is a view I never tire of.

Next I visit each horse with curry comb and brush. Had there been any concerns about sore shoulders or harness rubbings from the day before, I would take this time to check on their status. But this morning all is okay. As they eat I brush down each one in preparation for harnessing. Consumed as they are with eating I don't expect them to pay much attention to the naturally pleasant sensations of the morning brushing. If I should notice that one of them stops eating and acts as though particularly intense relief or discomfort is directly associated with a certain spot on their body I am going over, I pay attention, especially if it is anywhere that the collar makes contact.

My goal is always to get the horses fit and keep them in the field working. That means paying close attention to their comfort and care.

Next the harness: With no exceptions, each regular member of the working lineup has his or her own harness and collar. Today all the horses in the barn are mature and have worked long enough that I do not expect significant changes, day-to-day, in their collar fit. Sometimes, fat horses early in the working season will go down one to three collar sizes (inches) within a month to month and a half of hard work. Their necks carry a significant percentage of excess weight and as they sweat and work off that weight, the neck becomes thinner and shorter in depth.

There are many aspects of the harness horse system which are important, even critical. The fit of the collar rises to the top of that list. I pay more attention at the beginning of the season, but even now I frequently check to see how my collars are fitting. I want a perfect fit. If the collar is too tight, it will choke the horse down as he pulls and he'll quit from lack of breath. If, on the other hand, the collar is too loose (by a little or by a lot) the horse may continue working and you won't discover a problem until a sore has formed and the horse is in pain when it pulls. The damage from ill-fitting collars comes, 90% of the time, from the collar that is too loose. And it can be tricky because, sometimes, a collar may look and feel on the standing horse to fit perfectly YET when that horse pushes forward it gets either too big or too tight. When I find a collar that fits my horse I make sure I remember where I put it because it's the first one I want to return to when I harness up.

I've decided that this morning I will be working the two teams of mares first. The plan is to take the young ones out and open a hay land (in our case a strip 150' wide by a quarter mile long). I'll be going through a low area where there might be some standing water and tough mowing. These girls, I know, will keep the pace exactly where I want it, when I want it. The evening before, on the better mower, I sharpened the sickle, hit the grease zercs and filled the oil jug. It's ready. But I'm off the subject, back to the job at hand, harnessing...

In the tack room I get down their two collars, each with sweat pads fastened in. I run the flat of my hand over the inside of each sweat pad feeling for anything sharp or aggravating. I take one collar in the stall along the left side of the left horse. Leaving the collar fastened, I unsnap the mare's stall chain and slip the collar, right side up, past her head and down over her neck into place, careful to pull mane hair out of the way of the top seat of the collar. Then I fasten her chain back up. During this whole procedure this mare stands quietly and accepting, pausing from eating just long enough for me to do my job. I have known horses who would never put up with having the collar go over the head. I didn't raise and train them. Someone else did. And somewhere along the line they decided, out of fear or obstinance, that no one was ever going to put a collar over their head. In those cases, I unbuckle the collar and push it up from the under-side of the neck and refasten it topside, a procedure that is perhaps the safest bet for beginners but adds a half minute to my chores. And I like streamlining the process as much as possible. (I chuckle to myself realizing again that I seldom follow the rules and guidelines I have long given out to students at my work-shops.)

Next mare gets her collar on same way, nice and quiet, over the head.

I return to the tack room where the harness is hanging on two big spikes driven head high in the wall. I pull down the brichen assembly from the one spike, put it up on my right shoulder and run my right arm under it and down the underside middle of the harness till I grab low the right side hame in that same hand. Left hand takes the left hame about in the middle.

(All the miscellaneous straps, lines, bridle etc. have been attached, hung, tied, or fastened in such a way as to make my carrying the harness as uncomplicated as possible. And I do it the same way each time I remove a harness. It's very easy for me to tell when someone else has been dealing with my harness because things are out of place. When everything is in the place I want, this harnessing process goes quick and easy.)

I carry the harness out and approach the stabled mare from behind. "Get over honey," is my command. She should step to the right and up near her teammate instantly, but she doesn't this morning. "Get over!" I say with more emphasis, and she complies. Walking up on her left side, I lift the right hame high, pointing its bottom skyward, while pushing it and the harness up on her back. The hames go forward to seat in the rib-lined groove of the collar, with the connecting top hame strap at center top of the collar. Backing away slightly, I push each section of the harness, from off my shoulder and arm, up onto the waiting horse's back. Now it sits in a somewhat organized tangle atop the mare. Moving forward in the stall I ask her to back up and I pull the two hames to their seated positions on the collar. I'm careful at this point to see if I have accidently put a line or harness piece in under the hames. And I am also looking to make sure that the hames are properly positioned. They need to be equal on both sides, with the tug clip centered over the reinforced draft point of the collar. Everything is right, so I thread and tighten and buckle on the bottom hame strap.

The breast strap/pole strap assembly I prefer is removable, which means it fastens on both ends via heavy snaps. It's hanging fastened on the left side, so that when I throw the harness over the waiting horse, there is one less piece to concern myself with. At this point in the process I snap the right side of the breast strap to the bottom hame ring of the right hame. I leave the pole strap to hang for a second while I go back pulling and straightening the harness, gathering the brichen back over and under the tail. I check to see if the belly band in hanging straight down from the tug on the right side. Now I go forward, gather the pole strap between the front legs while reaching under for the belly band. The belly band goes over the pole strap and buckles in loose. Hanging from the two ends of the brichen are adjustable quarter straps with snaps on the ends. I fasten these to the ring at the end of the pole strap. My horse is harnessed.

I repeat the process for the second mare. To read back over the process description, it seems complicated, however, I have repeatedly timed myself and when all things are as they should be, it is simple and harnessing one horse takes between one and two minutes.

With the next, older pair of mares, there is a slight deviation in routine. One mare's neck is relatively small for her bulk, and her head is quite large, with lots of width at the eyes. This means that, though she's perfectly willing to let me try, it is close to impossible to put the collar on over the head (whether rightside up or upside down). So for her, I unbuckle the collar at the top, unclip the sweat pad from one side and pull it out of the way. I then pass the collar up at the neck, bringing the pad over and into place before buckling the collar together. The remainder of the harnessing routine remains the same as with the previous team.

The third team takes a little longer this morning because I need to find a better fitting collar for the one horse. I didn't like the way it rolled up on his neck yesterday. So I try a couple of collars on him that have a different shape, though they are the same size. One has been stretched out wider and the other is a full-sweeney style specifically designed for a thick neck. I find the right collar, a good older one, with a thicker overall construction and it requires that I lengthen out the top hame strap. That done, I proceed with the harnessing until all eight head are outfitted.

Next I separate the difficult mare to a single stall before I take out the young mare team. I don't want to come back later to the barn and find her gelding teammate bunged up and something broken, all because she slipped once again into her 'get away *from me!*' attitude.

Though there is a time and place for a single stall, I like the open double tie stall. It gives me a great deal of convenience. For example, I can drive a harnessed team directly into the stall when returning from work. And I can bridle, rig the lines, set the check reins, and back the team out of the stall when it's time to go to work.

Whenever possible, I work my horses without halters. Many teamsters prefer to leave the halters on under the bridles. I think this must add some discomfort to the horse and so I take them off whenever feasible. Mares ready, we back up and swing into the barn lane and walk out towards the waiting mower.

I walk the right mare over the mower tongue and swing the pair into place. They are standing now, either side of the tongue, exactly where they need to be for hitching. I walk alongside the left mare and spread the two lines across her back where they will be easily reached if needed. Then I proceed to the front of the team and raise up the tongue and neckyoke (the neckyoke is secured to the end of the mower tongue). I snap the breaststrap/pole strap assemblies to the neck yoke and take one quick look over the horse's heads to see if everything is okay. Next I walk back around the left side, picking up the driving lines and, with them in hand, proceed to hook the trace chains to the single trees. All this while the horses are standing calm, quiet and attentive. (The 'attentive' is important because I have found the inattentive horse is the one likeliest to jump when surprised, shocked or spooked by some unexpected sound or occurrence.)

All hitched, I climb aboard the mower, gather up the slack in my lines and check the team's ears. I want them both listening to me. Usually, the little vibrations, as I gather the lines, will tell them something's coming or that I'm getting set to ask them to go. Feeling my preparations, through the lines, can on occasion cause the horses to second guess me, and make them want to start before I'm ready. Seems natural, good intelligent working partners sense you are ready to set out, why wait for a formal command, why not step right out? Nope, don't ever let that happen.

They will remember it and take charge and that's bad news, that's the beginning of unraveled. It is my contention that we train our horses every moment we work with them. If we forfeit the opportunity to say, and thereby control, exactly when our horses step forward, we 'train' them to go whenever 'they' wish. A dangerous precedent. And one so easily avoided. It starts by **always** insisting *'we go when I say so and not before'*.

My young mares stand quiet, attentive, assured, ready. I smile and feel my breath shorten deep in my chest, it's not apprehension or fear or any negative reaction, it's that tightening that comes as a prelude to the waves of unavoidable natural gratefulness. I am so fortunate in my partnership with these beautiful creatures.

I gaze around til my focus returns. I give the command, *"okay ladies, let's do some mowing,"* chuckling to remember those hundreds of times I have admonished my workshop students to keep the verbal commands to their horses simple. And to always use the same sound or word for each desired action. *Don't do as I do, do as I say*. It's simple business and it's a complex craft.

We walk out to the hayfield and to that spot where, yester-day, I had tied a flag to the fence on my side. Also yesterday, I had paced off a new hay land marker across the field a quarter mile away, and propped a long stick in the cross fence. From the flag marker I am able to see across the relatively flat field to the stick. I swing the team in place with an effort to have my back be lined up, center, with the flag marker.

Sighting down the mower tongue, between my horses, I look over the 1/4 mile wide hayfield to the fenceline on the other side and line the marker stick up to a third point on the horizon, a tall tree. From experience I know that getting a straight first cut will depend on my keeping the distant stick and the far distant tree lined up and in my sight at all times. If I simply aim at the distant stick, my cut will wander.

Points spotted, I do a quick look over the team and lines, lower the cutter bar from the carrier rod and squirt oil over its length from the oil jug. Next I climb on the seat, lever the bar down to mowing position and kick back the pedal to put the mower in gear. Checking to see if ears are back, I speak to the team and we head out mowing. Smooth, quiet, certain. Everything is as it should be.

It's not always this way. I know first hand that mowing can be a frightening procedure for man and beast, especially if either or both are unaccustomed. I've been doing it for over thirty years now and this particular team has been at it for five years. We make it look simple and safe. And that is how it can be. But I always worry that first-timers will get the wrong idea from such a relaxed picture and jump into certain hazard or disaster. The picture must be earned.

We're mowing at a brisk walk, about 3 miles an hour, which is my preference. Even with these McD high gear #9 mowers timed, tuned and sharpened properly, going too slow can cause plugging in certain fine, wet, and/or wiry grasses. This is also why, when I'm opening a new land, possibly with wet lodged low spots, I prefer a team that will respond to my commands to walk faster AND be willing and able to stand quietly for a long stretch if I should need to clear a plugged cutter bar, or do some field mechanicing.

The team, on this opening pass, is walking through standing hay, belly deep. They love this run because frequently there is grazing available at a comfortable nose height. They've learned, over time, that if they keep a steady no nonsense straight ahead pace they can steal mouthfuls of grass and legumes as we mow. Fact is, I happen to know that they honestly think they are outsmarting me. They think I can't see. They think I don't know they're stealing bites. If they should stop or slow down to grab a bite, I would scold them. So they move along perfectly, trying not to let on that they are beautiful, clever thieves. As easy as they are on the lines, and with the snaffle bits, chewing and walking is no stretch for them. Our syncopation is built in small part on comic tolerance.

At 3 miles per hour we cross the field in 5 minutes. I never look back while mowing, keeping my eyes fixed on those two distant points until halfway when I've picked out a third point midway between the others. As I get closer to the end, I leave off looking at the stick in the fence and just concentrate on the mid-point and the far distant tree. Not until we reach the fence, do I stop and allow myself to look back.

Wooee, is that pretty! Straight as an arrow, mown hay laying back in a combed and symmetrical pattern. Feels mighty fine. I don't call this working, I call this making art.

I turn left, cutter bar towards the fenceline, and mow the 150' wide end and lift the bar, still in gear. We step straight ahead a short distance across the previous land, where yesterday's hay is waiting for the morning dew to pass (I hope to rake this hay in the afternoon). When the new mown hay has vibrated off the cutter bar and while we're still moving ahead, I kick the pedal forward and take the mower out of gear. We swing around on a U-turn to head back from where we came.

That first cut opened a new long hay land, a strip of hay up against previous strips, all part of a forty acre field. Each land is approximately 4 acres. I work this way deliberately rather than dropping the entire forty acres at one time. This allows me to mow four acres, next to another four acre land that I am raking which is adjacent to yet another land where the hay is being picked up or buck raked. I get better quality hay this way and fret a whole lot less about losing the whole field to weather or other uncertainties. If anything gets damaged or lost, it's usually just four acres. It also works very well with the horses as motive power. In fact work horses brought me to these sorts of conclusions, led me to thinking of patterns of working that give me the best chance of comfort and success.

Same trajectory, opposite direction, when we get to the land edge I aim the team to walk over hay they just mowed. I'm using the foot pedal to hold up the cutter bar. We stop just before the cut edge. I lower the bar and put the mower in gear. I speak to the team and we head off across the short end of the land. Twenty feet in I can see a ball up on the bar and that we aren't making a full cut. I tell the team to stop, take the mower out of gear, and use the lever to raise up the cutter bar. I can see what looks like a big nest plugged on the ends of two center rock guards. Off the mower and with team lines in hand, I clear the ball of nest and hay off the bar and run my fingers over the tops of the guards. I feel and hear something. Going back over with my fingers, I find a loose guard. I tie a half hitch of the lines onto the lifting lever and retrieve my crescent wrench from the tool box. After the guard gets tightened, I oil up the cutter bar again. Usually don't do this until I've made a few rounds, but I'm down and it's handy right now.

Mower fixed, I take the lines in hand and, speaking to them, walk around the left side to the heads of the mares. I offer them each a handful of the new mown hay as I spot check their bridles and harness from the front. I look up and around to make sure no one is looking and I plant a kiss on each soft nose. If anyone did happen to see me do it, I would deny it ever happened. Gives the wrong impression, to other people that is.

Back on the mower, we cut the remainder of the land end and I make a clean corner and head up the long side. The opening pass, having been made in the opposite direction, has lain the hay down in such a way that I can expect, on this cut, a couple of ballups and sure enough, one comes straight away. The inside heal has gathered a knot of hay and we're missing a strip. I stop the mower and this time I back up the mares just a foot before I raise the cutter bar. Usually this will clear the knot without my having to get off the mower. It works and we set out again.

We mow for two hours and drop more than half of the land before we head back towards the barn. I have more horses than I need, and some of them are out of shape. They need time in the field. This team could keep going all day and drop up to ten acres of hay if I pushed them. But no need to. I'm gonna go back and get another team to mow with.

I drive the mower over by the barn and point the team away from the barn door. I'm careful to park in a spot where the mower won't roll either direction when I unhook. The easiest way to tell this is to stop and back up just a hair to see if, with the tugs slack, the mower remains put. I get off the seat and unhook one tug when I notice someone driving up the driveway. I hook the tug back up and wait til the visitor gets out of his pickup.

"Cool! Those are Clydesdales aren't they?"

My Belgian mares roll their heads and shrug their shoulders both letting out a deep sigh. The visitor wants to help me with the horses and I tell him firmly that he can talk to me all he wants but I must insist he not stand in front of the horses or touch anything while I unhook. We visit while I unhook the traces from the evener and move, lines in hand to the left side. I drop the lines on the ground and go round to unhook the heavy tongue from the breast straps. I'm careful to hold up the tongue until completely free and then let it down slow to avoid hitting their legs or hooves. I unhook the check reins and go back to my driving lines. We continue to talk. The mares think it's time to walk off and because I'm talking, they catch me off guard. They succeeded in walking three steps ahead, so I stop them and calmly start up again and walk them around, they think they are going in the barn, but I turn them and we walk over the tongue and I make them stand as though we are going to hook back up. I talk with the stranger for ten minutes until the one mare lets loose with a long squeaky methane blast that raises her tail in the air.

"Sorry girls, yes it is time to go to the barn." And we walk off quietly, with confidence and comfort and with the whole of our lives musically defined.

Chapter Sixteen

Withstood
proof against the weight

It was warm for a March Saturday, warm except for the crisp intermittent breeze. I started the day by filling two temporary water troughs for the pasturing cattle and horses. The tanks we normally use are filled from the irrigation well but a Raven had committed suicide atop the power pole and the resulting short circuit had fried an electrical transformer. The power company said it would be a couple of weeks before we got juice back to that pump. So we're using a yard hose today to fill the temporary tanks. As is commonly, if wryly, observed, "that's farming."

After taking hay to the shut-ins, three horses, I took Lucky, the Australian Shepherd, and my fence tools up to where the night marauding elk had destroyed a corner of the fence. They would be back and tear it down again. And I will return to patch it again until we can figure out why the elk's migration has been disrupted, holding them in and near the ranch all around the year. Last summer one hundred elk ate half the hay crop. It was too expensive for us. One of many problems which cascaded down over us from late summer through to early this year.

I'm thinking about this later in the day as I walk across the 80 acre hay field retrieving pieces of broken wheel-line irrigation pipe which came loose of its moorings and rolled, pushed by winds up to 50 mph, until it hit trees, power poles and/or the cross fence. It will take time and money to repair it all. The cost of doing business.

But I can't think too hard on it because it's such a beautiful day and the soft lift of the awakening soil is a joy under foot. The smell says it's time to work the soil. I get down on hands and knees and comb the fallow ground with my fingers. I ball up the sandy loam and open my hand shaking it to feel the grains fall between my fingers. We are most fortunate to have our time on this splendid tract of mysterious fertile stony desert dirt. Tired from moving the pipe, I head to the house for a sandwich and a cup of coffee.

The next job on my list is to fix the jack on the front of my flatbed trailer. Seems something must have broke for it to fall off its blocking. I'll need this trailer to begin moving stuff to the rodeo grounds as our auction and swap meet is in a little more than a month from this writing. The trailer has some old rotten hay bales piled on it that I'll have to remove to a good spot.

At the trailer I find that the jack is okay and from the tracks, I figure out that elk have been crowding around the trailer to chew on the bad bales, probably the first time back in January when we were snowed in. They might have rubbed it off its jack block. It's an easy job to use the handyman jack and get the trailer up where I can hook it to the back of my pickup truck. I am thankful that in this case it's not the big problem I had originally thought it would be.

The breeze kicks up and I lift my head to it and smile when I feel the cold shaft of air whistle through the gap in my teeth. Again, as with countless times before, I thank goodness that I am alive and my eyesight is intact. I remember through a rush of returning physical sensations to the Fall and that confusing day of emergency surgery.

(Then I skip further back to last August and remember how, while fixing a hay field fence, I woke up face down in the grass without a clue as to what happened and how I got there.)

After the surgery, the doctor said I must have had that abcessed root canal for at least seven or eight years for the problem to have traveled so far in, through and past the bone. He said he was surprised I could still see, that I felt no pain, that I was still alive. He was not surprised that I had passed out. I had thought for months that the stiffening of the muscles around the eye and up the right side of my face was just part of aging. I had no idea.

Three months after the operation, the extracted tooth, the mouth full of stitches, the bone graft and the empty bank account, the surgeon pronounced the procedure a resounding success. I feel great, broke but great, far better than I felt for months before the procedure and purely glad to be alive.

I haul the trailer with the bad bales to the north shore of what we call 'desolation pond', named that because livestock grazing has stripped its banks of any vegetation. I spread the flakes of moldy hay in a thick mulch along the upper bank where I intend to plant grasses, cattails and shrubs as a habitat belt for wildlife. While there I rethink how the proposed fence will go, thinking that a wide spot around the old Juniper tree might be encircled by poplars and shrubs and in that way make a good viewing blind for watching the snow geese, the merganzers, the mallards, the bald eagles, the herons, the mule deer, and the coots. I'm excited about the project and the discussions we are having with our neighbors about forming the non-profit organization to formalize what has been happening on the ranch towards wildlife habitat enhancement and applied research.

Larry and Sue drive by on their daily round to their quail test sites, where they will check the traps and put out more feed. Don and Jim called yesterday to say that it was time to put out some new Kestrel hawk nest boxes and would Scout and Kristi like to go along.

    I unhook the trailer near the shop and go inside to work on a disc harrow for the auction. Even when it's a mess, like today, I get comfort from my shop, the tools, the projects waiting so patiently for my time. The space is a small safe haven, not as intensely personal but much the same as my painting studios are. And as anxieties mount, with this last season's financial, physical, family, and weather-related struggles stripping us of the hopes we had for winter farm preparations, I need safe havens. I need those direct uncomplicated, mind settling little jobs which offer up small rewards upon completion.

    Weather inviting, I take a chair from the shop and move it outside to sit and work on a tangled pile of baling twine. The shop building shelters me from the cold breeze and the sun hits me square. The warmth is what I like to think of as a drawing warmth. I see the horses and the cattle standing and on the ground, muscles limp, soaking in the penetrating rays. The dogs are spread out flat on their sides, noses pointed away from the body, legs spread, so that a maximum area is exposed to the sun. The peacocks are spread out as well, not for the warmth but to take advantage of how the early spring sun makes their magnificent tails shimmer. When the cold is invasive, the animals and humans hunker down turtle-like, shoulders up and limbs close in. With the first fresh warmth of early spring we mammals unfurl ourselves and stretch out much like the emerging new plant. The warmth draws us out from our winter curl and huddle.

    I am sitting on an old metal folding chair, leaning back with cap cocked even further back on sweaty hair and brow, legs stretched out forward as far as they can go. I pull loose one cut length of baling twine at a time and tie an end of it by square knot to another length. In this way I am making one long twine I might use for various purposes, perhaps to trellis beans or peas in the garden. The warm sunlight, the birds spring chatter, the resting muscles, the resting brain, the slow methodical finger work are calming. I am able in this moment to reflect on the recent past and to project thoughts forward to how we'd like to see the near future. And all of it is a model for what our difficult days most require. Calm reflection, positive outlook and steady small steps forward. Often easier thought than done.

I think about how the true economy of the nation has put a tourniquet on the budgets of so many friends, out of work - less work - more financial pressures. We are told that inflation is at a standstill but that doesn't ring true when gasoline hits 2+ dollars a gallon, steel & lumber prices soar, and electricity bills keep climbing. Most of us struggle to get by on less. I wish I had the means today to help my son as he manages valiantly his new difficult world of single dad. He'll make it, he's strong and two year old Eva is the ultimate incentive. We work at finding creative ways to help them. I smile a familiar father's smile when my thoughts move on to Juliet and her many successes with theater and career. And I chuckle as my twine fixing reverie is inter-rupted by a scream of "watch this one Dad!"

Scout goes whizzing by on her bicycle, shoulders down low and forward face up flat to the oncoming air, arms bent up at the elbows bug-like, as her little legs pump the pedals for all they are worth. She navigates a quick turn around a puddle in the dirt road, manages a tight figure eight, and just when it seems she has pulled off her trick, falls over with her bicycle into a clump of grass. The look on her face is one of surprised disappointment. She gets up slow and offers "I'm okay." Two children in their late twenties and one not yet ten and me better cut out to be an uncle than a father, a friend rather than a guardian, a playmate rather than a model.

She walks her bike over to me. "That was a good trick honey, sure looked scary, did you hurt yourself?"

"Yeah, see." She points to a red elbow.

"Ouch."

"Whaddaya doin Pop?"

"Untangling this string and tieing it into one long string."

"How come? Don't you usually just throw it away?"

"Today, honey, I feel like sitting here and making something useful from this pile."

"Oh...   Watch this one!"

and she races off, elbow forgotten, to threaten the laws of physics one more time.

As I watch her, my face swells up with gratitude and I think how fortunate we are to be living here right now. I think about the challenges of millions of children and families in Ruwanda, Iraq, the Gaza strip, the ghettos of Baltimore, Afghanistan, and countless other struggling corners of the world. I think about fathers whose children have been killed or maimed. I think about children whose fathers have disappeared forever. We are fortunate to live here, but I am conflicted by my worries that we are dangerously complacent, we are so disconnected from the terrible suffering of millions, perhaps billions of fellow humans.

I shake my head and look straight up at the blue sky struggling to focus. We've had a bad winter but it was nothing, NOTHING, compared to what so many have withstood. And that is the word which shifts me and my perspective, WITH-STOOD.

I like the word and its suggested meanings and I promise myself to look it up in the dictionary later.

I reach for a short piece of discarded dowelling and I tie the end of my string to it. I then wind the string into a ball on the stick, alternating the wrap to amuse myself. My thoughts shift to a quote I have hanging on my computer screen at the office.

"Ask that your own life have a meaning," Albert Camus

I looked up the word and found this limited definition in my old dictionary:

Withstand. *To stand against; to oppose or resist, with either physical or moral force; specif., to be proof against the weight, pressure, influence, etc., of, as, to withstand the force of a storm; to withstand ill fortune.*

I was somewhat disappointed because I had begun to feel the word differently in its parts and possibilities. "With" and "Stand." When we have *withstood* I feel as though **we** have prevailed, **survived**, succeeded, rather than to have, in a competitive sense, won against something or someone. I am drawn to the part about being *proof against the weight*. We Millers have withstood this challenging Winter.

Even with a positive outlook it is highly possible, within a life and a living, that there are spans of time, or short spells, when a person or a family can feel buried by a pile up of unexpected, unpleasant, tedious, difficult, threatening and destructive events. Buried and alone. Challenges which can bring a person or a family to its knees, challenges which work to destroy any positive outlook. For some of us it may seem it's the rule rather than the exception. But all things change, and we must see the better future. We must have faith that better is always possible. Let other people in to walk alongside.

Farming is an enterprise which rides on faith. It is an ingredient in that difficult to explain mix of motivation and appreciation which permits the hope of farming. We go on blind faith that there will be a spring which is conducive to field work, we go on blind faith that moisture and sunlight will combine in good measure to bring us our crops. For those of us who are farmers, but of course for all human beings, so much depends on a positive view of the possible and a belief in our inherent abilities to be a beautiful living proof against the weight.

I am startled from my weighty reverie by my daughter's words.

"Dad, did you see it!? I went all the way up past Roo's pen and up that rock! My bike was in the air! And I didn't crash! Did you see it?"

"Honey, You are incredible. I am sorry I missed seeing it with my eyes but I could sure see it inside my mind as you described it. Let me try it once?"

"Okay, but be careful Dad, it's scary and you have to hit that flat rock just right..."

Please excuse me folks, I need to go prove myself against this new challenge.

Chapter Seventeen

making it better

All through our lives there are parallels. Often the story of one work, one adventure, runs parallel in meaning, value, and substance to other overlapping stories.

pipes

We irrigate with sprinklers and pipe. Have done so for 16 years. It takes one mile of sprinkler pipe, 128 + or - sprinklers to cover 120 acres in haphazard fashion. Three quarter-mile wheel lines and one quarter-mile of handline. Wheel lines, in our country, feature a large aluminum wheel clamped to the center of each length of pipe which has a sprinkler on one end. This pipe is clamped at both ends to other wheeled sections working out from both ends of the tractored-center. What you see is a 1300 foot long stretch of aluminum pipe with a big wheel every 40 feet or so and a four-wheeled frame at the center with a small gas engine mounted on it. That long wheel line is connected, in our case, by a flex hose to a ten inch diameter aluminum main-line pipe

which runs at right angle to the wheel lines and supplies, on
demand, pressurized water through gated valves. Two electric
pumps are used, one 100 hp three phase pump to pull the water
out of the glacier-fed underground lake and push it into an upper
lagoon, and the other to push the water out of the lagoon, and
pressurize it down through the mainline. It takes a great deal of
electricity to run those pumps all summer long.

To some "experts" this system is one of the best ways to
deliver exact amounts of water to field and crop. Who is to say
what "best" means in this case. Our system was installed in the
mid 60's and, if I can keep it running another few years, it should
be eligible for the historical register. The idea that this *Rube
Goldburgian* mangle of pipes should, after forty years, provide
service, wrecks havoc with depreciation tables and the hairy
axioms of efficiency experts.

This system, when it is up and going, is theoretically easy to
run, it requires that the pipes be moved every 12 hours to a new
setting. This requires that I shut down the valve for the individual
wheel line, disconnect it, drain it, start up the little motor and
"walk" or roll the line 60 feet to the next valve. There it is recon-
nected and turned back on. I watch to make sure that no "birds"
or sprinklers need attention and that the line "seats" - meaning all
the gaskets and overflows have closed. Then I move on to each of
the other two wheelines. After all three are moved, I walk to the
handline, shut the valve, and one pipe at a time move it sixty feet
over to the next setting, reconnect and turn back on. With no
wind, no breakdowns, best of all functioning worlds, I can get all
the pipes changed in one hour but it is more reasonable to allow
an hour and a half. Combined, that comes to 3 hours out of
every day.

This year, through the winter and early spring, we have had
harder, steadier, long lasting winds than in the past. These winds
had torn into the secured staked-out wheel lines and ripped
everything loose. Steel posts used to secure the lines, were twisted
into the wheels. Nylon rope snapped, and aluminum pipe twisted
and tore itself apart. And then sections of liberated wheel line

rolled, pushed by the winds, across the fields, over barb wire fences, to wrap bent around Pines and Junipers.

I was familiar with each little piece of damage, as I had dealt with these things over the years. I, however, had no working concept of the total mass of the problem. All I could do was to set in and work on each individual problem until my system was back up and running. Because of the urgencies of Journal work, amplified by implementing exciting plans for new additions and changes here, I found myself challenged to get the irrigation system fixed. Frustrations set in. But I knew, from years of trying to be a farmer, that all I could do was keep at it.

There was the day when I thought, maybe, just maybe, I would be able to turn the system on. I pressurized the mainline and looked down its half mile length at fifty pin holes shooting out sprays of water. I shut it down and called a young man with a portable heliarc welding outfit. What we discovered is that the mainline, running alongside the cross fence, had been used as a pathway for the big herd of nocturnal elk trespassers. They have danced on the old aluminum pipe poking little "v" shaped dents into the pipe with several of the dents resulting in pin holes. He welded what we found and left. When I was next able to pressur- ize the line, I found fifty more holes. First thought was to replace the line but that met with a choking laugh as it cost too much. I hunted for and found four different tube products which all claimed to work to repair aluminum pipe. One did and amaz- ingly well. But I had to test each and allow time for them to set up.

When finally, late for the season, I got the irrigation run- ning, should I say limping along, I felt a quieting cautious sense of accomplishment. Made it... for now. I thought back a little bit and realized that, when I finally resigned myself to it being a long slow repetitive process, I had come to enjoy what I was doing. It was actually liberating in a moronic way. Walk down there, shut that off, walk back here lift this, roll that, replace that, plug that, walk back there, open that, watch and decide. Think for a little about what was next and repeat the whole routine with a slightly different twist.

If finances allow, I think I have made a critical decision about this irrigation business. I want to change it dramatically. I want to research setting up a flood irrigation grid off, or out of, a set of three lagoons, all of which might be used to raise trout somehow protected from the Eagles and Osprey. If we do this, I realize that I will be leaving behind the routines of the sprinkler pipes. I know if I do that the working rhythm, with frustrations and successes, will remain within me a part of my makeup just as dance and commercial fishing and carpentry have stayed with me, not as skills but as working rhythmic memories. My body still remembers, even after all these years, what it means to "nail off" sheeting or decking, what it means to "pull rigging," what it means to repeat a dance step a million times. In the same way, I know I will walk down that length of irrigation pipe for the rest of my life, and into the next lives of my families. I am a better person because of my experience with the pipes, it has branded me. I am thankful for the civilizing influence of my pipe fixing odysseys.

gratitudes

We enjoy work. We also enjoy rest and celebration. For farmers and ranchers Thanksgiving, by definition, seems a most appropriate celebration. But it does not come often enough and we owe it to ourselves to feel and share our thankfulness frequently. It is in this way that we remind ourselves of the depth of our lives, the value of our living, and the force of fertility.

Just down the road from us a few miles, folks gathered at Bob and Gayle's ranch recently for the annual pre-Father's day pre-rodeo barbeque. Children, grandchildren, neighbors and friends from far and wide descended, literally, down into the steep, secluded, lovely, sudden, deep creek canyon of the Rimrock Ranch. On the banks of the crystal clear tributary of the Deschutes River, beneath the Ponderosa Pines and Cottonwoods, a scattered encampment of tents lay framed by green meadow and windy blue sky. Everybody brought something but each of those somethings, warmly welcomed as they were, did little but provide trim to the amazing repast which tumbled from boxes, coolers, cookers and grills. Gayle and Bob provided fresh oysters as appetizers, followed by a large deep tray heaped with steamed clams, mussels, and crayfish, laced with wedges of lemon and lime. As family and friends reacquainted themselves and took drink and shell fish, *John of the Latitudes* labored over the large adjustable grill to prepare for the quail, corn and buffalo. *Texas Tom* and *Vern the Venerable* tasted tidbits to protect the women and children from themselves while I made certain my shells, bones and leavings were regularly and surreptitiously transferred to other plates. It was my self-assigned duty to make the lighter eaters appear more robust and I took that responsibility to heart. I know that if I am to have any thanks or credit for that small and dangerous work it will have to be self assigned. So I thank myself here and I also confess that the food was fabulous which made my trials bearable. But that is getting ahead of the story. The main deal of the meal was of course the buffalo and quail along with beautiful salads, roasted corn ears, excellent breads, and exotic beers and wines. The eating melded with various conversations from hunting yarns, to travel notes, to gardening desires, to ranching challenges, to the vagaries of television programming, to fashion as design versus attitude, to bird sightings, and finally to the difficulties of producing network coverage of the Grecian Olympics. It was a various group of people and so the next course, that of dessert, needed to match that variety and it did. Strawberry pie, a frosted adorned carrot cake, sweet lemon bars, a tray full of brownies and exotic cookies, a subtle soft fruit cake, and more. All culminating in bodies crowding a campfire for warmth as the sun went somewhere else and the evening breeze turned crisp. The coming darkness slowed the children and softened the conversation as eyes reflected the fire and suggested secrets remembered. Guitar music and singing gently sent people on their way, each and everyone feeling well filled by amply-shared life-defining thankfulness.

That celebration of gratitude at Bob and Gayle's filled me

with a sense that we can be open to our lives being thick and various, we can let in the civilizing influences, it's our choice. And that, as with many things I think about, led me to thoughts about the work we choose and the shape we allow that work to take.

thickness

We can have our farming thick or thin, deep or shallow, full or empty, intriguing or depleting, fertile or sterile. In many respects, it's our choice. We can take a modest chunk of money, today, for whatever we can scrape off the topsoil in the shortest, easiest and most convenient time frame possible. Or we can share in a deep wealth of fertility, health, promise and energy, a wealth of *long-lived, regardless the years*. We can make money nervous and on the run, or we can make money dancing. We can greet each day angry with our insistence on no frills, no inefficiencies, no romance, no poetry, only a serious pursuit of a professional moneymaking agriculture. We can farm smiling about choices made and trees planted all of which will bear fruit long after our time. We can tear out the fences, fencerows, and old buildings that are doing no good except to harbor birds and varmits or we can design pasture fencing and time our forage harvest with a view towards that land's future fertility, birds with homes, and our real profit. We can plant a crop with a view towards what it will lend a crop as yet unplanted. We can select a breed and a sire with a view towards three generations hence. We may graft fruit trees out of long range curiosity and discover enchantment. We may enjoy generations of 'cooperative' customers who through their appreciation and connection, add intangible beneficial aspects to our deep fertility. We may be reminded, each time we look at a tree we planted or a favored old animal, of a friend from another time. We may look upon the remains of a dirty, hot, bloody process and marvel at our own abilities and resilience, our own undeniable relevance. We may ultimately look across the landscape of years of our efforts and see both our imprint on it all and our perfect insignificance. We can, in that experience, feel carried aloft and cradled down deep.

We have a picture in our heads of how our farming is or is to be. The ingredients, livestock - crops - operations, are pieces of that picture, not as numbers or profits or equations or assets, but as shapes, colors, textures, emotions, in a mosaic of motivations. And we, each of us, and us together, are also pieces of that picture. The picture is not of our making so much as it becomes us or is becoming of us. We do look good in the picture. The picture is incomplete or nonexistent without us. Yet never lose sight of the fact that we are also unnecessary for the picture.

Going way back and far forward. Or sitting on a thin spot and in a short period of time. Farming, all tied up with and tied into everything which has occurred for years, decades, generations on this place, in this soil, with these families of people and animals. Our self-definition is cumulative. Over time we become part of who we are, just part. What riches are ours in how we see who we are and what we do! What true wealth there is in seeing and enjoying our good fortune! We do have a choice in how we see it all. And how we see it shapes what it is.

citizenship
Yet, our wider world is far from all it might be. And we must for our own mental survival never forget our citizenship.

Today, for so many in so many places, it is a dark time to rival the worst in man's history. A shrinking number of people indecently controlling a huge percentage of the wealth while hundreds of millions struggle with great courage and decency to hang on to life. So many wars, conflicts, threatened clashes, disasters, and new deadly diseases that we can, out of emotional survival, only allow ourselves a quick view of the most dramatic. Two dozen people, somewhere on the planet, dying of hunger every minute of every day, while hundreds of millions of dollars are gathered and spent to get someone elected to federal office so that they might promote and protect the assets and growing wealth of the people who paid to put them there. Big business squashing tradition, heritage, genetics, families, dreams, soul accounts, and soil banking. To fight them, if we run in like some

Don Quixote, noble and comically refreshing as that may seem, we waste the energy of our good will. It is challenging, especially for those who feel they MUST do something, but we have to see that working at what we do and who we are, in the most unselfish and regenerative ways, IS helping to make the world a better place.

Each rush to get the first piece of pie, each grab, shove, and act of hording, feeds the monstrous social evils of the day. Each time we pass by a fellow human being who is stranded or hurt, we feed the evils. And our western society encourages the behavior. We have institutionalized our problem. We have shut out the civilizing influences.

There is a pervasive and insidious argument across the land which regularly makes the case for the right and good benefits of *competition* in a *free global market*. Definitions of free market set aside for the moment: these arguments set forth, as vulgar notion, that the clearest indication of success comes from how competition would cleanly determine the best places for a given commodity to be produced. As if there is some higher law at work.

They say: Apples which have long been grown in key areas of North America cannot reasonably compete with cheaper apples grown in Asia ergo *give it up man*. They say: Beef cattle which have known a long and profitable tradition in North America struggle to compete with that which is grown in and imported from Oceania and South America. A hallmark of North America in general and the U.S. in particular has been the fact that our culture and the natural resource base allowed for a remarkable and limitless capacity for self-sufficiency. We could, and have, produced all of our own food and fiber up until recently. Now the elected powers have us shutting down our farms and factories in favor of purchasing from other countries. There is a chance of a big paycheck today from the sell-off of our self-sufficiency. It comes from multi-national corporations who stake claims on the labor pools and natural resource base of third world countries. These corporations pay our elected officials, bureaucrats, and commercial scientists for their complicity and they dare to call it patriotic, they dare to suggest it is for the good of us all.

In the shallow fevers of these difficult times, when politicos take advantage of our collective emotions for their gain and power, it is so dangerously easy to confuse the calls to nationalized rationalizing pride. I try to remind myself of the way pride can function like a drying agent, shriveling substance to a shadow of its former self.

I prefer the heart wrenching soul searching pride of the late sweet Ray Charles singing *America Sweet America* with the meaning of the words riding up on top of the melodies and the emotions. I want the civilizing influences for me and mine, for us.

Bruce

In 1971 I was old enough to think I was old enough. Today I am young enough to see just how old I was back then. Most of my false maturity was carried on the shoulders of critically defined formative hungers and attractions. With two or three years of mixed and mixed up involvement in farming, I had just enough of a taste to know it was the life for me. And my imagination was working overtime to shape the future of my involvement. I knew what sort of farm I wanted. I constantly drew out maps and designs of that farm of my future. And those pictures always included a place for work animals.

That year I took a job managing a small ranch for a man who unwittingly shaped my life by allowing one of my most important beginnings and by showing me, over all these years, the power and value of supportive natures. From the first time we met, Bruce Leonhardy believed in me, even when others around him ridiculed and pointed out obvious flaws in my character. Bruce mixed encouragement with a genuine fascination in my interests and growing abilities. Because of his keen interest in me, he learned early on of all the trappings of my farm dreams. He listened to, questioned and marveled at the intensity of my interest in diversity and scale issues with farming. He was curious about how I managed to construct beneficial relationships between separate aspects of a dreamed-for farming operation. He doubted, with the patience of one who cares, that certain things would actually work. Yet he never doubted me. He wondered how I would pull any of it off, and he wondered how I did it when eventually it happened. Yet, in perfect contradiction, he also never faltered in showing his belief that I could do it all. Now I can look back and clearly see how intoxicating and valuable it was to have this person, unrelated and without responsiblity to me, believing that I could do it all. And he was just as ready to be surprised and pleased when it actually happened.

I had learned of Belgian horses being offered at auction on a farm and mentioned it to Bruce. He handed me a check and said, *see if you can get that dream team of yours for this, I know you'll pay me back some day.* He had an impulsively generous spirit which pushed him to invest in people and ideas that tickled him. Later I would learn just how much.

I went to that auction, in a borrowed truck pulling a borrowed trailer. And I bought my first team of horses, Goldie and Queenie. And, as the old saw goes, the rest is a rush of small history. It would take me over seven years but I eventually surprised Bruce with an envelope of cash to repay him for the loan plus interest. He cried. Turns out his fortunes had reversed and the money came at a good time for him. Also, it turns out that Bruce had loaned thousands of dollars to many individuals and, at that time, I was the only one who had repaid him. He had moments of depression when he felt no one believed in him. So, for me to show him what his support had meant was particularly poignant for both of us.

As time passed and I gathered small notoriety for my work with the big horses, with the books, with the magazine, and with my artwork, Bruce would puff up to delightful proportions recalling how it was that he had made an investment in the future of a young man who had gone on to do things of some note. He'd say "I knew you could do it," and "compadre, wow, you did it!"

Recently Bruce passed away from us, me. And along with

the great sadness that comes from loss, I felt gratitude for the immense generosity, support and friendship of this gentle, curious, and especially comfortable human being, gratitude for having shared some life and dreams with him. For all of the nearly thirty-five years that I have known him, Bruce has always been perfectly disguised as an ordinary fellow. If you let him in close to you, you knew he was pure spirit in thoughts, in deeds, and in the music of his manner.

There are important lessons in my relationship with Bruce Leonhardy, lessons which I have tried to use with the work I do on *Small Farmer's Journal* and the books I have written. I have repeatedly used words which come down to one large sentiment directed towards farmers, "We believe in you, we need you to succeed with your dreams, and that you can do these things." Under it all, I have understood, through my own failures and successes, how incredibly important it is to feel the support and encouragement of others when struggling in the dark with unfamiliar efforts, untested waters, challenging goals, and difficult turns.

Just as with my old irrigation system, the *Small Farmer's Journal* adventure has gone through some challenging times where the only thing to do has been to keep going, keep working, fixing each little thing as we get to it. We are warm and certain in our conviction that this is not a business about profits, or prophets, neither is it a non-profit influence peddlar, it is a small community as rich and varied as that celebration meal at the Baker's ranch. Neither 'either' nor 'or', it is something else again. A baldfaced long-lived effort to say that the civilizing influences of shared stories, recipes, building plans, decent procedures, natural selections, manure teas, and dirty broken fingernails belong together, that's what the SFJ effort is.

How do we know when a rhetorical question, one which appears to make a statement rather than elicit a response, falls off the fence to one side or the other demanding true consideration?

For example, if we choose to keep it the same, how long will our old irrigation system last? If we modify that irrigation system within our own values, will we realize some improvement? If we keep close to home and ignore the political furors and the war cries, how long will our value system last? If I go hunting for windmills and old armor to wear as I strive to make political points, will we realize any improvement? If we work hard to protect the wildlife habitats which are possible in our farming world, will the rest of the world be better served? If I hold close the pervasive power of genuine supportive natures such as that of my dear friend Bruce, will the world be better served? Or is it time to question our values and kick apart the clumps? I used to think the future was ours to make of as we will. I now suspect we were the future, once, and that the jury is still out on how we did. I am going to fight that suspicion and keep doing it all, only better.

Chapter Eighteen

Deeper Yet

I'm an old guy, my ways and ideas should be set by now. Have you ever had a few consecutive days that adjusted your attitude and outlook? This summer I had a couple of weekends that taught me things about patterns, influence, economics, politics, humor, and success.

They were not big weekends, not momentous in any way. They were quiet, restful, working, shared weekends. Started with a road trip: up, down and across the rolling hills of eastern Oregon. That road trip set a rhythm to my observations and altered my perspective slightly.

I had planned to write an essay about influence. What part influence sometimes plays in the development and growth of an individual farmer. How we influence others without really trying. How we pick our influences and how they pick us. Mighty important.

And then I thought it would be better to speak of the economic and political climate we find ourselves immersed in. How many of us are having to come up with our own new working definitions of frugality and thrift. Of how frightened some of us are with the turn of politics and the divisions we feel within our communities. And yet how tremendously hopeful many agricultural trends are, how thrilling many people's farming adventures are, how connected and connecting are the myriad triumphs we see in the rural marketplace. I thought of telling funny stories about farming and its challenges. I thought about sharing the details of highly successful small farm adventures I have recently heard of and seen.

The Road to Dufur

These thoughts and options were all swimming around in my brain when and while I found myself driving through eastern Oregon on my way to the annual Dufur Threshing Bee. It was their thirty-third year, I had not been to the event in thirty years. Kristi and Scout had never been. We wanted to lend our small support to this important, somewhat local, farm process re-enactment. (It takes two and a half hours for us to drive to Dufur, a distance and a time frame that we out West consider *somewhat local.*)

There are Threshing Bee re-enactments and festivals scattered all over North America. For those who don't know what I am speaking of, a *Threshing Bee,* in days of yore, was an event where neighbors gathered together to help each other separate harvested grains. Way back when, a well-equipped farmer, or an individual subcontractor with a steam tractor and threshing machine (two major pieces of equipment), would travel from farm to farm to separate the grain, a job which required many hands. Families would come together at each farm and make a *Bee* or working party of the chore. Modern events such as the Dufur Threshing Bee have been organized and staged to demonstrate, through a re-creation, the old equipment, horsemanship and individual skills (i.e. sack sewing, shocking, bundle loading, etc.) as well as the entire work party aspect. The events are staged to keep the traditions alive and to educate folks about the heritage. Some of these demonstrations are more tractor than horse related, or vice versa. But they are all guided by a common goal of sharing the actual workings of a colorful and important tradition. I like to think that the highest value of these events is that they keep **alive** the working knowledge, the skills, the techniques, the secrets, the balancing act. When we put the information about such procedures, no matter how well illustrated, into books and magazines, we do a service but it can never match the vitality, complexity, attraction, and information of seeing and possibly participating in the **real** thing. Being there you realize that *it goes deeper.*

The first time I attended the Dufur Threshing Bee, the event was just three years old. Way back then I wasn't paying attention to such details, but in attendance, helping master teamster and header operator Ray Guthrie, was a ten year old boy named Mike McIntosh. Today he is the master teamster in charge of the Dufur Header with youngsters helping him and learning. But, once again I'm getting ahead of myself.

It was a hot August day, pushing mid-nineties on the high dry-air desert. I was driving a pickup without air conditioning. I was pulling our old travel trailer and a load of books and magazines for our little event tent booth. We were running late. Kristi and Scout had doubled back to the ranch because lightning strikes had started twenty small fires on and near our property. The fire fighters were on the scene but Kristi had to go back to restart our irrigation system so that they could load their fire tenders without having to drive an hour's round trip for water. I went ahead to try to get to the event on time.

The last half of the road to Dufur, old highway 197, is up and down and sideways for stretched miles which seem longer than normal. There are two lovely, close-in, visual breaks to the endless high desert sage and wheat fields. One is when the road drops sharply to the Deschutes canyon and the little river-rafting mecca of Maupin. Today this town is bustling with tourists all either just out of or just going on to the river. The old town absorbs them nicely but without membership. These visitors to

Maupin look out of place, out of their element, passing through, uncommitted.

Pulling the slow winding grade up from the river bottom, I drive by modest abandoned hundred year old dark red and silvered grey pine-boarded buildings, specific in their once-upon-a-time utility and now as much a part of the landscape as the rocks and the sagebrush. And I also drive by newer mobile home influenced architecture of the deliberately affordable type. Some of these are perched precariously on the side hill, perhaps to give the best possible view of the river below. The old woman on that porch is surely the sweetest thing I've seen in quite a while, talking to the local, they're probably going to share some food later. And, sitting on the curb is a very modern teary-eyed young woman who has just broken up with someone. The man approaching her, flip flops on his feet, baggy low hanging flowered shorts, and a bandana tied round and over his head - he's a city boy here for a hot weekend of rafting and partying.

Like that Jerry Jeff Walker train song, this town, she's so easy to read, it's so easy to peg the visitors and the locals. Their stories, transparent, are right out front to read. But when the evidence arrives, I find out I was wrong about most all of it. The flashy visitor turns out to be a local and the local turns out to be a lost salesman on the road to Madras. The heart broke girl turns out to be happy as a clam, her visage was artificial, that mascara-enhanced photo-ready artificial power-angst that youth simply must project. And the sweet old lady is nothing of the sort, she's bitter and sarcastic and vengeful. The hundred year old weathered pine shed was built two years ago out of used materials to house the raku supplies for the local Polish potter. What do I know, what information can I trust, maybe it goes deeper? None of it could be true. Should I care? What's more important, my first impression fitting snugly into my demanding expectation? Or the truth of it all?

But on that long slow hot road, the theater of passing through the town was lovely diversion.

Up top again for another long haul until I drop down into

the pristine little landscape of Tygh Valley. The rolling yellow wheat fields and dusty saged bunch grass cascade down to the dark green willow and cottonwood trees which stretch out, meandering and sheltering, on either side of the creek. It's easy to imagine kids sitting in the shade and fishing for rainbows. A scene more reminiscent of the early landscapes of Northern California's Napa Valley, back before easy money and rock hard new-found sophistication fell into viticulture. Grapes and wine have become synonymous with BMWs, plastic surgery and a dominance over the landscape.

Reminded me of the lost enchantment of Jack London's *Beauty Ranch* in Napa, and all of the dream farm gymnastics he cultivated. His Shire horses, his rock piggery, the ingenious covered manure-tea still, the cultivated Eucalyptus and Cactus groves. He poured his writing profits, from books like *Call of the Wild*, into that horsepowered organic dream decades before the rest of us were old enough to even imagine such magic. And he did it in a way that came after the landscape, not upon it or in spite of it. Of his adventure there, he made a life more song than working.

Funny where your mind can take you. I see Tygh Valley and conjure up like views, London's ranch - my home place, and rationales for wanting all of it to remain a secret, happy to see many people drive the highway oblivious, anxious to get some place *real*, some place else.

Dufur Threshing Bee

I pull the grade out of Tygh Valley watching for the little sign that will point out the left-turn side-road to the town of Dufur. Coming from the south, you can't see the town from the highway. It is protected from view by the rolling hills. I turn left and wind into the town's edge at the exact moment that the Threshing Bee parade is being set up. I manage to join, curb side, with friend Jim Jensen, of Oxbow Trading Company, to watch what's to become one of my more treasured parade experiences. I have been in several parades and have watched many more over my six decades, but none that served me up such an appropriate small, understated, living poignancy as this one. Maybe I was open to it; jaded, tired, exhausted by the current anxieties, perhaps I had a hole in me that only a genuine, sincere, small, poetic, comic procession could fill. Rags and jewels, stomping feet, pets, horses, shiny new and old tractors, wagons, toys, steam whistling out of the steam tractor, beautiful side-saddled woman in flowing hat and skirts atop a fine-boned zebra dun mule of perfect proportions, a dusty dirty old model-T convertible with bicycle and boxes strapped to its sides and a grinning whiskered old guy who seemed to enjoy a private joke that grew as he wheeled forward. I imagined, easily, that this fellow in the old car had just been passing through town and the parade had swallowed him up and carried him along, in between shiny classic cars and Nascar wanna-bees. He was enjoying the folks on the curbside, waving to him in generous appreciation of his existence. Then came someone I know, little Jacob McIntosh ground driving his family's big team of Belgian geldings with papa Mike following behind to make sure the lines didn't drag on the pavement. No band, no majorettes, no politicians in cadillac convertibles, no pretense of any kind, no corporate advertisements or sponsorships, just a sweet old-fashioned down-the-middle of the street parade, in a small, still proud and elegant little American town, nestled in its own hollow, amidst the rolling open wheat hills a very long ways from the rest of the world. Lovely.

As the parade wound down, I went to set up our little book booth on the town's curbside and was helped by my buddy Ed Joseph. He had shown up with his family to take in the weekend. As we set up the tent and unloaded boxes we had a brief conversation about challenges he was having, or had had, with his team of work horses; a conversation laced with his desire to get more happening with his working horses and farming dream. It felt good to see in Ed the future. I kept the sentiment to myself, it can kill a conversation. Each positive look into the future warms me, I feel it like a puff of air lifting us on. But that's not something you want to say out loud.

Kristi and Scout arrived and set off to take photos of the field demonstrations while I sat at the booth visiting with folks. Directly across the street from where I sat, a family sliced chunks of watermelon and cantaloupe to sell to passing people. Just to the left of them, and adjacent to the entrance to the threshing area, sat a seventies vintage Ford station wagon with shaving cream writing on its windows. *'For sale, project car, runs, $75'.* Everybody who went into the demonstration area had to pass that car. To the left of our booth a stationary baler sat as sentinel to the historical museum behind us. It attracted many people of various ages who either bragged at having a working knowledge of it or wondered after its purpose. Further down a dozen or so craft booths displayed their wares.

It was a hot late morning, sun shining and smells wafting up from the food vendors. Then I saw her for the first time. Maybe fourteen years old, black hair cut in an odd angular fashion atop a head whose sole purpose seemed to be as display-hanger for as many piercings and rings as possible. Attractive girl with an extreme tatoo on her upper arm, a cut-off tank top and low hanging jeans which together seemed to push out her slight belly in an exposed fashionable roll. What drew my attention to her was the fact that she was walking down the middle of the road, head up, proud as can be, just as if she were walking in the parade? But the parade was over. Cars had to go around her. She made me smile. It was after all her town, no doubt quiet most of the year, and strangers had come to town this weekend to be entertained. She, by golly, was going to be seen.

At that point, five people all zeroed in on the Journal booth. One man, who looked to be 35, tapped a cover of the Journal with one finger as he nodded his head in approval, recognition, or both. The man and woman with him bent over looking at book and Journal covers with a projected sense of ownership. One of the men wore a Portland Farmer's Market cap.

But it was the young man, maybe 19 - blond - tall - angular - poised - polite but insistent, who drew my full attention. He stood back a few feet and looked from book to Journal to book as if to say 'what is this?' I spoke to him offering an invitation to look and a short description of what the Journal was about.

"I'm not a farmer, I live in the city, but I care about these issues," he said. "What's the best way to support these issues, the farmers?"

The question caught me a little off guard because it was so direct and, in a way, generous. I couldn't help but sense other questions just under the surface.

"The quickest, easiest, most direct way to help is to think about how and where you buy your food and make positive changes there," I offered.

"I don't understand." He answered and I noticed that the other folks still perusing the booth had become keenly interested in the discussion. He told me he was from Portland and I offered that he should visit the city's farmer's market and take time to talk to the vendors, take time to ask about the produce and the farms.

"If you start buying direct from certain farmers, either at the market or from the farm, the food will mean more to you, it will taste different because it is fresher but also because you will put flavor in from your new found knowledge and familiarity," I added.

"That's right," chimed in one of the other men, "you should check out the Sauvie's Island doings. That's the best scene..." And he went on to share several particulars as I moved sideways to sell a couple of books to another visitor to the booth. At that moment I heard my name in a *hello*. Looking up I greeted old friend Mike Johnson driving an ATV which pulled a long chain-train of barrel cars. Each little car was made of a poly barrel set sideways on a fabricated wheeled axle. A rounded opening section was cut in the topside of the horizontal barrels allowing children, one each, to sit in the cars and be pulled around. It looked all the world like an old fashioned carnival ride that had escaped its moorings.

When I came back around to my young friend, he looked disjointed and awkward. I tried to change the subject and asked him about his life and plans. He was in college and trying to

decide what his future might be. He was here at the Threshing
Bee because something itched, something pulled at him, he didn't
know quite what it was. Our conversation came back around to
this first issue of how he might help small farmers, and I began to
tell him something of the phenomenon of farm produce subscription sales or what are sometimes called CSAs. He seemed to
tighten in on the idea as he gently fingered the current issue of
the Journal. I looked around, we were alone for the moment. I
handed him that Journal. "I want you to take this copy as a gift."

"No, I couldn't do that."

"Yes, you can, because I'm confident that you will read part
of it and want to come back and buy an issue or a subscription.
So you see I'm not being generous, I'm being sneaky."

He thanked me, smiled and turned to walk away towards
the field demonstrations. He didn't get far. He came back and
said with some difficulty, "You see, what I really want is to find
out if I could be a farmer. If farming is something a person could
do today. I mean, is it reasonable? After all, I would want to be a
good farmer. Thank you again." And with that he walked away.

I had not said a word to the last set of statements and
questions. I couldn't. I felt like I had witnessed something private
and fragile. I had witnessed this young person's life choice. Not
discovery but choice. He hadn't signed up or enlisted or volunteered yet, he had simply chosen. He had finally seen something
in the near distance that had given him permission to follow an
attraction. His statement, "I would want to be a good farmer,"
reverberated. He wasn't talking about being a successful farmer in
any profit-oriented model. He was talking about being good for
farming, good at farming, good because of farming, good with
farming. The more I thought about it, the more emotional I got.
I had been present at, and a small participant in, the beginning
visions of a new farmer.

A trickle of folks visited the booth and looked and sometimes purchased items. The talk was light and undemanding. I
was arranged emotionally to observe and respond. Every ten
minutes or so a horsedrawn wagon, either Joyce Sharp or Dave

Peterson or the Harmans clopped by with a load of relaxed happy people. Time passed slow, refreshing and thick like a cold fruit nectar. Friends had come by the booth to give us tickets for the steak dinner in the park that evening.

Then she came back again, this time from the other direction and again right down the middle of the street. Same head up, eyes averted, strut. No shuffle here. Only this time she wore a pretty, sleeveless, patterned cotton dress. There seemed to be less jewelry and her hair was brushed back wet. It was the same young teenager transformed. But still on display in her own parade.

A woman thumbed through a book on wheelwrighting when the first siren blew. Her husband rushed to her side in anxious whispers. She said to me, "I'll be back, I want to get this book for him, but they just told us that maybe the wheat next to our house is on fire." She rushed off.

Kristi came to the booth to relieve me. Scout, she said, was swimming in the creek with the McIntosh kids. Would I check on her? I went, with her camera, to check on our daughter and towards the threshing area to visit Mike McIntosh. He had suggested that I might try his six Belgians in the field demonstration on the two bottom plow. More sirens went off and the polished fire trucks, parked in display after the parade, peeled out with lights and warning horns as a bearded man directed them on to the street. I went up to him, he was crying.

"Is there anything I can do to help?" I asked.

With his good arm he kept directing traffic. As he spoke I could see that one side of his face was contorted by nature and that the arm and hand of that same side curved tight to the body and was difficult to manipulate. His words were slurred.

"The fire's up by my friend's house. I should be going with them." He was overcome with his limitations, with concern and uselessness.

I took a few steps and stopped, looking up to the near horizon where a column of smoke was visible. All around me, most people were continuing with the event and the day's doings. Looking back at the crying man, I felt odd, sad and odd. I walked past the many craft demonstrations and the Threshing machine lost in thought and blind to the goings on. Worrying about the wheat field fire set me to worrying about our own lightning started fires back home and somehow caused me to focus on worrying about our nine year old daughter swimming in the creek.

At the creek I had to chuckle to find the half dozen kids, my daughter included, stomping their feet in water that could not have been more than six inches deep. A few feet to their left, under a canopy of Willows and Cottonwood, someone held the lead rope on a draft horse which drank with hesitation.

Out at the field I watched as Rocky Hegele, Jeb Michaelson and Josh Kezele plowed. Thirty years before, Jeb and I had both been in attendance at Dufur when a universal joint went out on my pickup. He stayed behind a while and fixed it to get me on the road again. We go way back, so to speak. Today he, and his fine team, strike a very strong figure on the walking plow and header box.

Everett Metzentine, TyghValley rancher and legendary wheelwright, had been instrumental in getting the Dufur Bee started thirty-three years hence. On this day he was talking to folks, explaining all the various aspects of the field demonstrations. Mike McIntosh drove his six Belgians on the plow over by Everett for the explanations. Later, on the other side of the field, Mike handed me the lines and I got a chance with his wonderful four and five year beauties. The McIntosh family shows their horses around the state in six and eight horse hitches. They also demonstrate every year at Dufur.

Yard Sale Race

I believe it was later that evening, sitting and swapping stories with the entire McIntosh contingent, in the center court formed by their circled travel trailers, that I first got an inkling of what I am certain will become a new American sensation, a competitive sport so daring, so exciting, that Olympic status is most definitely on the horizon. You need a little more background on the Dufur parade and community in order to understand this amazing development.

Dufur is a small town which is all but taken over by the yearly Threshing Bee. Many people come from neighboring towns and the city of Portland. Recognizing the golden opportunity, a few town's people set up garage and yard sales during the weekend. This had grown to be community wide. The biggest concentration of visitors, spread throughout Dufur, is usually during the parade. The parade makes a circuitous route through town, past many of these sales.

As the history of this new sport goes, it seems one of the McIntosh group, Cameron to be exact, had, a year or so ago, been in the back of a wagon during the Dufur parade when he spied, ahead, a yard sale. He bailed off the wagon and made a beeline to that yard while Mac warned him that they wouldn't be able to wait for him, horses and wagon would be continuing on. Cam hollered "I'll be right back, keep going." Then he did an amazing thing. He managed to make a deal on a vacuum cleaner WITH attachments, pay for it, gather it up and get back on the wagon, without the wagon having slowed or stopped. The parade did not skip a single beat. Come to find out that he had done it many more times, "scoring" as he likes to say, many great deals. Now he counts on the parade as a unique shopping opportunity. We all laughed comfortably with the tale.

I could see the serious future of it clearly. To all of them, Carnahan, Clay, Cam, Becky, Josh, Mac, Mike, Joanna, and Kristi, it was a funny tale to tell. To me it was obviously the beginning of something big, really BIG. So I sketched out to them how we could develop this into a formal rule-structured competition that might be played out across America. There would be members of the local Historical Museum Staff which would need to be drafted as judges. Contestants would be restricted in this manner: once a yard or garage sale was spotted along the route, they would have to call out their "target" item, jump off the parade vehicle, purchase the item, and get back on the vehicle without it stopping. They would do this repeatedly with a goal of having not only the most items but also the best deals and an exhibition of courage matched with guile and bargaining skills. When the parade concluded, the judges would tally the items, grade their relative quality and the "deal" making, and mix in any eye witness accounts of bravery and derring do, and come up with a winner who would receive a trophy looking suspiciously like one of the yard sale signs. Scoring would be standardized so that winners from Dufur could be measured against winners from other parts of the country for a national champion. The only thing lacking in the plan was a name that measured up to the entire concept. What would we call this sport? Yarding? Scavenging Race? Jump Off Racing? I still don't know why they all laughed at my idea. I think it could work and perhaps develop into a team sport.

Word came to us that the wheat field fire of that afternoon ended nicely. The owner of the field had simply turned on the irrigation system and snuffed it out immediately. We went to our travel trailer and bunked down for the night. It had been a good day.

Next morning sweet Joanna and happy Mike, made everybody breakfast. Then we went together to the church service.

"Red wire blue wire, red wire blue wire, which one do I cut? If you see me running, you'd better get a move on." These words repeated bracketed the bearded preacher's Sunday morning service in the tent. The post script came when the lovely older pump organ player in bonnet and apron asked if she could take a photograph of us all for their scrapbook.

This day I was able to take in a little of the crafting displays which included two different rope making displays, one with our old friend John Reser of Condon, blacksmithing, wheelwrighting, and a bunch more. Mike had suggested he might let me try my hand on the Header and Kristi wanted to take more pictures, so Cam and Becky volunteered to run our booth for a little while.

It was late morning and just before I set off to visit the Header I was surprised to see the girl again. I shouldn't have been. But this time she had an entourage. Again down the middle of the street, in shorts and T shirt, bejeweled and with some sort of glistening wax in her now spiked hair, she sported an entirely new upper arm tatoo! I wondered for half a second if I was confusing this apparition with the girl from yesterday but there was no mistaking the posture. And this time she was followed by a line of younger children marching along as if to a "follow the leader" rhythm. As she parted traffic, I noticed that the cars were all waxed and shiny classic hot rods entering town from the city of The Dalles, here for a display at the city park. The visual juxtaposition of this young lady against the flat-topped sixty year old hot rodders was a delight. Pure Americana.

I walked slowly down through the display area heading out to the field demonstrations. It was hotter than the previous day, classic dry high desert summer heat punctuated by smells and sounds. One sound was particularly sweet. In the tent where church service had been there now played a wind quintet doing old summer show tunes. Peaking through the corner of the tent I thought I could see a lovely older woman wrapped up in the glistening pipes of a French horn. It was fine, mighty fine.

I made it back to the booth during a quiet time, it was afternoon and things were winding down. A little ways away I noticed the young man from the day before. He was intent, arms wrapped around, looking down. When he caught my eye he came over.

"Did you get a chance to read anything?"

"Yes, I did. I really had no idea. So much to think about."

"My name's Lynn," I offered.

"Mine is Joseph." (Two Josephs in one weekend, Ed Joseph and now this young man.) He picked up another Journal and handed me some money, it was more than the price by a couple of single dollar bills. When I said so he remarked with a smile that he wanted me to have it *all*. I had the distinct impression that it was all the money he had at the moment. I was once again speechless.

I looked up and around, feeling good and wanting to remember Dufur this way. I was hoping to seal the deal by seeing that young girl marching down the center of the street one last time, but it was not to be. I'll just have to come back next year, enter the yard sale race, and sit down in the tent to listen to the wind quintet at the Dufur Threshing Bee.

Appreciating Reminders

I said at the beginning of this writing that I had been thinking about the subjects of *influence, economics, politics, frugality, thrift, and farming adventures,* but chose instead to share the story of my weekends. At this point in my life I am more interested in the actual workings than in the rationales, apologies, or theories. I'm looking for the good stuff. I believe the stories of our lives can give us a useful and constructive view of economics and politics. Economy is more than making money, it's about balance and true value. Good economy is about the folks at Dufur selling slices of watermelon, the girl claiming the middle of the street, the old Ford for sale, yard sales, children playing in the creek, and a work party helping to get the oats to market. Good economy is generations working together with the pure power of happy accomplishment. And good politics is **not** about politicians lying to us and jerking our lives around and scaring us into voting a certain way. It's about people finding collaborative ways of sharing and helping one another. Good politics is about the physically impaired man helping to direct the fire trucks. Good politics is the strikingly handsome form of Jeb Michaelson comfortable on the handles of his walking plow. Good politics is the country preacher reaching for poetry to share a conviction. The best politics is about the volunteer wind quintet playing music in the tent on a summer afternoon in a small town that refuses to die.

For me it's about keeping my eyes, ears, nose and heart open while I'm working. All about observations; how we see what we see, what definitions we assign to those observations, what we allow ourselves to discover on down the road, and what we learn or unlearn from our observations. It's about going back beyond. It's about going deeper yet...

Chapter Nineteen

Improving Farm Income

"We think of life as a solid and are haunted when time tells us it is fluid." Jim Harrison - The Road Home

Ever notice that when you blow in a dog's face he gets mad at you, but when you put him in the car or pickup and take him for a ride, he joyfully sticks his nose in the wind?

Many of us are good farmers (at least with the growing part) yet we fail at our business adventure because we are somehow crippled when it comes to seeing how our produce fits in the world. And seeing how our produce fits in the world is the front porch to feeling better about ourselves and realizing greater farm income.

Some things we just don't talk about. How much we charge, how many we sell, what it costs us to produce, how much money we have left over, how much money you have in the bank, how well we're doing, what our farm is worth on the market?

Many of us, me included, are mixed up idiots when it comes to questions of income and finance. We prefer to keep our money affairs to ourselves. We don't tell (or teach) our children about it. We frequently keep such matters from our spouses. And we darn sure don't want the neighbors to know the truth. That goes for success, failure, and the inbetweens.

In many parts of the country, in many of our regional sub-cultures (for example up-state New York, Chicano southwest, Louisiana bayou, Ohio coal country, Seattle suburbs, southern Indiana Amish country, Maine woods, Vermont interior, North Dakota exterior, etc.), these questions of farm income and personal wealth are left to speculation and the unspoken consensual assumptive measurement of external vestiges; new pickup, acres in tomatoes, number of cattle and horses, new house, painted fences, number of laborers, new clothes, haughty attitude. Together, we at some point in the distant past agreed that we should take these things to be clear proof of success. But we should know better. These things may be as much an indication of debt or flaunted inherited wealth or clever natural style or a short-lived self-destructive mind-set. It may be important to you that you appear successful, so you consciously or subconsciously adorn yourself to convey success. It may be important to you that you appear socially conscious and progressive, so you dress down for the party. How you are situated in life may be the result of a low self esteem which insists constantly that you not rise above your place in life, so your appearance becomes an apology. These are factors which gum up the works for us when we decide to work at increasing our farm income. These things are important to us, they do go a long ways to defining us. In so far as they might be debilitating and counterproductive, I need to believe we can change how we see ourselves because it may be essential to success.

It's a values thing. Not the often hypocritical family/religion/values thing which we saw bought and sold in the last national election. It's a truer indication of what we care most about, closer to the bone. And it works coming and going. How we see ourselves and how others see us. No amount of talk to the contrary will erase the obvious. If the president of a non-profit humanitarian aid organization lives in a million dollar home and drives luxury cars, whether he's up to his eyeballs in debt or not, it does cast a pompous tone over any notion of his having any unselfish posture of service to mankind. And vice versa, the crusty old coot who drives a broken down rusty pickup truck but has a net worth of several million dollars will be hard pressed to break out of the class assessment which comes with his appearance.

We kick it around so much that is has definitely become a thorny cliché, but it is still no less true that outlook and attitude can have a greater affect on the assessment of success and wealth than the actual measured pile of dough. It goes right to the heart of the formulaeic wholistic approach, where all things are measured. Whether inside a theoretical grid or not, farmers need to understand their needs structure and the effect of imagined goals. The most successful farmers usually extend such thinking to inquiries about what potential consumers need and wish for. Putting the two together can result in a powerful formula for success. That said, we still have to be realistic about the balance of wealth and how it affects who farms and who doesn't. We need to understand that there are forces out there poised to take from us any gain.

Most agribusiness, inextricably tied as it is to an unapologetic *savage capitalism,* cannot afford a concern for individual independent farmers, their families, their communities, food health and safety. And the long-term future of farming doesn't even show up on the radar for corporate-ruled agricultural industrialists except where market and resource control comes to play.

Bill Moyer warns us not to disregard the very real tension between the haves and the have-nots. He goes on to say, "*Civilization happens because we don't leave things to other people.*" In that direction I am keenly interested in what I know to be the civilizing effect of creative marketing coupled with a humanitarian value structure. I believe that when we, each of us, succeeds at marketing our good, clean farm produce, we become a civilizing influence on our society.

I have friends who get all itchy and uncomfortable when they hear me talk like this. "You gonna pull up that old 'class' thing again, the *haves* versus the *have-nots?*" Nope, that's not what this is about, not this time anyway. This is about food and farm income, and the crazy existential notion that we don't need to take anything away from anybody in order to balance the scales.

This is about recognizing that some excellent farmers are failing to realize the income <u>available</u> (not owed, not promised, but available) to them. And it's about breaking through some difficult and downright dangerous barriers to do the talking. This is a look at possibilities, it is not a formula, not a promise, not a road map. There is no guarantee that following a certain plan, any plan, will result in a measured increase or success.

In the privacy of your moment reading these words, it won't seem so threatening to hear that your position in the world needs a lift, needs affirming, needs jazzing up some. But truth is, farmers are generally held in low regard. Out in the open, the so-called real world will throw on you a flush of embarrassment and ridicule if you hazard to present yourself as an example of worthy vocation. And we need a lift in our self-assessment because there is a place for *influence* in the design of market approach. We do not sell or project our influence well if we accept a low measure of our vocational identity. How do we, as farmers, position ourselves to have maximum influence on or over the marketplace, the decision makers, etc.? We do it, in part, by believing in our value to society. For God's sake, we grow food and care for the earth! Does it get any better than that!?

what do we want?

We make a clear distinction between industrial agriculture and farming, between agribusiness and agrarianism. Agribusiness is about the extractive production of commodities. Farming is about the craft of growing and replenishment. We favor farming.

On the farm front, if frequency of query is any indication, the largest concerns folks have shake out this way:

1. How can we make a decent living farming?
2. How can we afford a farm of our own?
3. How do we learn all that we need to know about success-ful farming?

4. How do we afford converting to an all organic operation?

5. How do we guarantee that what we have built here, on this farm, has the best chance of continuing beyond our tenure?

Though these are typically personal concerns expressed by individuals and/or families, they are perhaps more importantly seen as true community concerns which go to the heart of the health of food and of the countryside.

And these questions do seem to circle around the one subject of *farm income.* We need the farm income to realize a decent living. Good farm income prospects will answer question number two, and direct priorities regarding number three. Good farm income would beg of question number four that better returns are not only a prime basis for conversion but make it possible to begin with. And question number five answers itself as well: an attractively profitable farm will be inviting to those who might carry her forward.

It is my contention that just as important as individual committment to improved farm income, rural communities MUST work to see their farms succeed and do well. There is no other long term solution to the health and vitality of the country-side, small towns and their institutions. Without a healthy fabric of independent farms and ranches there is little to keep small town North America alive. Sure, a handful of mining and logging concerns might glue together a few communities, but the larger share of the rural countryside is obviously empty without its farmers.

fitting in

Not too long ago, I was at a social gathering and met some folks who spoke fondly of several years of past experiences they had enjoyed operating a small mixed farm. It centered on a highly successful CSA (or subscriptions sales club) of produce and meat buyers. They had quit the enterprise and "moved on" to more profitable, if less enjoyable, 9-5 jobs. Their CSA customers, after a few years, still lament the loss of the farm. And this lovely, sincere, intelligent couple obviously missed the whole of the farm. But, as they said, they couldn't make a go of it, not enough income.

I asked what they produced and was surprised at the wide array of organic fruits and vegetables they made available over a long growing season. And in addition, they also raised meat chickens, eggs, lamb, wool and assorted other items. How much money did you make in a good year? I asked. No immediate answer. A mutual friend who sat in on the conversation offered, in their embarrassed presence, the observation that theirs *is* a beautiful farm and, when they were operating it, it was highly productive and obviously very successful, everybody thought so. Except for the operating couple.

I interrupted, you said their farm *IS* beautiful, I turned to them, do you still own it? Oh, yes, they said, we live there now. All we have growing at the time is an asparagus bed and some flowers. There followed assorted conversational reminiscences surrounding the information that they had inherited that farm, and that it was free and clear of debt. It was obvious they loved the place and their recent memories as full-time farmers. The adding machine in my journalist head was clicking away at what were interpretive numbers: x number of chickens, x number of eggs, x pounds of cut and wrapped lamb, x number of CSA customers each receiving x number of baskets of fresh produce, wool, fruit. I had to press on with questions.

"Do you ever think of going back to the farming?"

She looks over at him, he at her, she leans towards him and smiles, he looks down at his folded hands.

"No, well not really, it just wouldn't be realistic," she offers. When they quit farming, they both took lucrative jobs, she at a government position. Coincidently she had just lost that job the other day, lack of funding for the project, but was not thinking about going back to farming, much as she loved it. She would find another similar position. And he is locked on to a career path and, much as he also misses the farm, can't imagine quitting the lucrative and secure post.

"Do you like the work you do?" I asked him.

"It's okay, it's what I went to school for, I'm good at it. But, if you're asking how it compares to the farming we did, it doesn't. It can't. But that doesn't matter because we simply could not make a go of it."

At this evening gathering we had all shared a meal. Fabulous organic pastured chicken had been served at dinner which was produce of our host and hostess' own separate farming operation. In defense of the young couple who I was questioning, our hostess offered, "It really is no different for us. We wouldn't be doing this farming if we didn't believe in it. Because we aren't making any money. When we butcher chickens we have a regular bunch of customers who come and buy every one we have to offer. But we don't cover our costs."

"Then you must raise the price," I offered.

"We can't do that. We'd price ourselves right out of the market. Besides we have good friends as customers who couldn't afford an increase."

"How long will you be doing the chickens like this?" I asked.

"I don't know, we talk about it, we may have to quit doing it."

Others in the room chimed in 'you can't do that!'

I turned to the former farmers and asked straight out, "How much were you making with your farm?" After hemming and hawing they finally offered that in a good year they made a gross total of $40,000 and kept $24,000. It was not enough money for them to stay with it.

"By raising your prices, you could have been making twice to three times that much," I observed.

They didn't argue that point. Maybe they could have raised their prices but they wouldn't because, in their view, it would not be right. They weren't talking about temporarily keeping their prices low so that they could attract customers and build up a market base. They were talking about having lots of devoted customers and a waiting list of people who wanted to buy from them. And yet, out of that curious guilt of the new progressive American farmer, feeling that they had to sacrifice themselves at the altar of neo-communal capitalism, they gave away their beloved produce and ultimately, as a consequence, the farming way of life they loved. It was another example of selfless generosity as both strength and weakness, but in this case weakness won out.

Leaving names and locations out of the narrative, let's say that these folks owned and farmed 65 acres with ten acres in cane fruit, orchard, and market garden. The balance in pasture for sheep and chickens. I have to agree that, in their case, as successful highly productive farmers with an established "hungry" group of customers, 40 thousand is wholly inadequate by the standards of the last ten years. By USDA extension service standards they should have been realizing at least $75,000 gross from this model. And by "our" standards, their income might, or should, have been between $100 thousand and $300 thousand gross per year. What was the problem? Where is the missing link?

I suggest that, first and foremost, their prices were too low, and that they were giving away much of what they produced. Also that there were missed opportunities to add aspects and value which would have put product into an entirely different realm. (I am reminded of a conversation with Eliot Coleman of Maine and how he and Barbara have selected mixtures of vegetables suitable for stir-fry and gathered them together into sealed plastic bags of convenient portions. This relatively simple way of altering how the product is offered allows for a substantial increase in price and, with freshness and top quality, an ever increasing demand.)

But I must hazard a guess that, for our good and intelligent, now former farmers, had they been making three or four times more income off their farm, an issue of inadequacy would have persisted. In this day and age, how we fit into the wider society has greater weight than we are sometimes willing to admit. Construction workers, union and otherwise, are frequently paid more than government issue bureaucrats and white collar managers, yet the prestige in our society is usually afforded to those who are perceived to be in control. People regularly take lower paying jobs because they want to *trade up* to higher prestige. Perhaps, where it applies to farmers, this is something we bring upon ourselves by allowing and even encouraging a certain pigeon-holing. For example, the generally held public notion that any individual interested, today, in farming as a vocation can't be too bright because it's all hard work - low pay and even lower prestige. Once again I say they got it wrong. It is hard work, good hard work (that's a positive) and the pay can be outstanding. Plus, it doesn't get any better than being able to work at what you love, producing good food and fiber and all the while helping to keep your part of the world healthy and vibrant.

So, I am frequently asked, after I once again suggest that prices need to be raised, if I'm not encouraging small farmers to become greedy capitalists like the big boys. I am not suggesting that we abandon the humanitarian streak which seems to come so naturally to the independent farmer. To the contrary, we must protect it. Robin Hood was right. We have exceptional produce in high demand, but well-hidden by our low self-esteem. We need to let everyone know we have this valuable stuff for sale. We should sell it for all its worth and then apply some of the profits to return in the form of food to those unable to purchase that produce. If someone is truly unable to afford the pasture-raised poultry, that person should have an easy and honorable opportunity to trade for the food. If someone is able to afford the poultry and complains about the price being too high, we should scratch that one off our list of customers with a *'thank you very much, sorry it didn't work out, check with us later'*.

Before we go too far in that direction, something needs to be said about community economics and character. When this couple quit their farming venture, the immediate community suffered a loss: a loss of income, a loss of a piece of the local self-reliance, a loss of some of the invisible and intangible bracings that hold together folks who might otherwise not interact, and most of all loss through the diminishment of character. When they shut down, their community shrank. The loss was all the more pronounced because what they did with their farming was loved, appreciated, needed, identified with, influential, and beautiful. All this is to say that any community which stands by and watches while these farms, their islands of self-sufficiency, hope, flavor, heritage, and beneficial works, walk off for lack of support, is short-sighted at least and, in my view, doomed to a vacancy of spirit. The flip side is that those communities, and they are out there, who rally around their farmers and find creative ways to keep them in service and hold them as cherished, those communities large or small have vitality. And that's what it is all about, having vitality.

The measurements of success go in several directions. It is a trap to measure, in linear fashion, one income against another, especially when one of those incomes is derived from a working life which affords a vast array of wonderfully difficult to measure benefits. Each of us must come to grips with our own take on the difference between having good health and affording health care, of being happy in our work and working to be able to afford happiness. "Affordability" is the grand modern economic buga-boo. Vitality should be the goal.

who decides what the price is?

It's a broad subject with endless argument, but suffice it to say that how food is generally priced and, in a perverted way, how food is valued, does not derive from edict or law. Because your local supermarket chain sells its often dangerously stale, poison-ous chicken at such and such a price does not make it worth that much or that little. It is as much the result of wholesale industrial contracts, temporary marketing incentives, distance to market, corporate parlance, and dumping. Yes, it is true that people can keep shopping and eventually find any food item priced at rock bottom. But that is not the person, regardless of personal income level, who prizes freshness, cleanliness, taste, safety, and personal connection with his or her food. Who says a 5 pound pasture-raised organic chicken must sell at a certain price? What is it truly worth on the open market where people struggle to find clean, healthy food of true freshness?

It used to be a question of supply and demand, pure and simple. Not so any more. It's gotten complicated and disconnected.

Irony of ironies: in parts of the California interior, where thousands of acres of monoculturally produced fruits and vegetables are grown, large segments of that nearby rural population suffers from a malnutrition so advanced as to cause many preventable conditions and diseases. Good fresh fruits and vegetables are not available even at retail market prices. The produce is all shipped out to wholesalers around the country and world. And whatever remains in the field after harvest is guarded until it is plowed under. In the not so distant past, this area was the proverbial horn of plenty with a diverse array of small farms mixed in with the larger operations. Milk, cheese, meat, eggs, and produce were all readily available. Our society traded that, all in order to rip out the fences and produce a handful of crops by chemically intense methods. Imagine, people in California's land of plenty go hungry for lack of access to food.

Today, at risk of criminal charges and arrest, people are smuggling into New York City, raw milk and raw milk yogurt. Farmers are sneaking into Manhattan at night with coolers packed with this contraband. They are lugging their product upstairs into apartments where they are meeting secretive groups of mothers and fathers anxious to trade fists full of money for what they consider to be pure gold. The going rate today in this black market trade, is just under $5 a gallon for raw milk (not including deposits on the glass bottles). Such a scene of the blackmarketeering of banned foodstuffs you might expect of the old USSR, not today's US. But alas, under the disingenuous mantra of food safety, in my view a set of officially sanctioned lies as laws, more and more fresh fruits, herbs, meats, dairy products, etc. are being denied the public. We know that the practical realities of the industrial food storage/distribution/handling system which results in fast food, regularly poisons thousands of people, but NO effort is under foot to close down ANY aspect of that infrastructure. However, if one lady gets caught selling raw milk to human beings in New York City, the entire weight of the federal government comes crashing down to squash mode. Ah, but you see, this can be seen from a different angle as opportunity.

As reported in a recent New Yorker magazine issue, a very popular and successful Manhattan cheese shop, uniquely offering a vast array of obscure and magnificent cheeses from all over the U.S. and the world, advertised that they were interested in finding a dairy farmer, within commuting distance of New York city, who might be able to supply the store with an artisan grade of butter. They received NO replies. Their detective work indicated that there were no longer any small dairies within commuting distance of New York City *legally* making and selling real butter!

In such a world, I repeat, we should sell our farm products for more money and at the same time make sure they are available to anyone who wants them. We can and must do both.

how do we change things?

These are our challenges: first we must come to the successful growth of farm commodities. Next we need to identify our market universe and commodity pricing possibilities. We must never lose sight of destructive competition and always work intelligently to insure against encroachment and downright theft. It is part of our job to keep it alive. Then comes actual marketing: we must communicate with our customers about the safety of our food, the true amazing freshness, the available conveniences we are able and willling to offer, the exhiliration of the farming experience and how we might share slices of that with them. We need to make our products and farms more accessible: each farm needs to have its printed list of products and services with pricing, ready on demand. Wherever possible, we need to stay away from most conventional wholesale marketing and that includes brokers of organic foods. We must never steal from our enjoyment of the farming process and always build towards the future : plant and animal breeding - tree planting - perennials - farm design and evolution - all these things and more will eventually add to the profitability. We need to think creatively about converting expenses to assests and income. For example, think about allowing for work parties as a way to get customers to touch and feel the farm. Understand that the potential customer is wooed as much by the grace and beauty of the operation as they are by the food itelf. The best customers will be those who are candidates for strong personal identification with farming. Don't rule out those who insist on convenience; they will pay for it.

Learn to think about your prices with a core charge, or base line, and allow for ways to add value and thereby additional income. For example, determine what you need to have for a flat of eggs. Price it according to your costs, plus a reasonable return to you. Think retail, not wholesale. Now, if customers want to purchase a traditional carton of 12 eggs of a certain size ADD those aspects to the price. If they want to purchase a half dozen, ADD that aspect as an option. Allow your customers to read where they may be able to purchase the flat of eggs at a reduced price per egg and the additional cost for the convenience will be theirs to accept or reject. Write this all down on a sheet or brochure that you can give to the prospective customer. Make sure the brochure includes clear statements as to your farming methods, the quality of the produce, and the FRESHNESS! Think about including in that literature that, for cash strapped folks, you accept trades and are willing to offer special deals.

For many of us, these direct marketing ideas don't help much, but I want to argue that they do point towards an attitude and approach that may be customized to fit many situations. And keep in mind that wherever we are able to back off the industrial model of farming; the model that has us produce a lot of a few things and sell it in bulk to middle men, wherever we can back off that and find ways to sell direct to our community and region, our chances for profitability soar. And better yet, our own sense of place in our immediate world turns to gold.

In our computerized world there are myriad ways to get the word out about our farm produce. I want to offer a caution that we don't make the mistake of accepting the dangerous illusion of it being a small world and that our market universe can include Tasmania, Brazil, East Cameroon, and South Korea. We need to keep our own local community and region in clear focus. Belonging here does pay dividends to family, spirit, land, and history. It is a big world, vast and terrible and exciting and unknowable. We are small pieces in that world. Keeping it in close to home, holding focus on who we are and what we do, that is the way to success and vitality. It is about improving our farm income but never at the expense of who we are. Because who we are may be the thing that saves this beautiful planet.

For God's sake, we grow food and care for the earth! Vitality! Does it get any better than that!? LRM

Chapter Twenty

Back to Basics - We know these things.

Fall 2005

On one of my many trips early this year, a couple shared a story with me of their farming adventure. It was a remarkable tale of a handful of cows, a modest market garden, and a couple of draft horses resulting in income, natural connections, and a serenity that surprised this couple. In the beginning, it was thought they had retired, by choice, to hard work and uncertainty. Instead they found themselves wonderfully engaged in daily efforts they loved, profitable efforts which gave them a cloak of assurance rather than repeated slaps of uncertainty. They saw their story as small and insignificant rather than exceptional. Watching their modest faces and seeing there the genuine thrill they felt, sharing particulars with someone whom they suspected appreciated their good fortune, I was smiling through the moment. It was a story I knew I wanted to hang on to, possibly reference in my talks and writings, and definitely squeeze for the juice it continues to offer my own farm dreams.

Then came a phone call from another couple in another part of the country. They were calling to renew their subscription and had caught me on the other end of the line. We visited for a couple of minutes and I asked about their farming adventure.

"Oh, it's not much," she said. "We have eight cows we milk. A couple of Percherons to do the work on our twenty-three acres. We sell raw milk, cheese, some vegetables. We couldn't have done it without the Journal, the information and stories have been so inspirational. And, of course, our two daughters, we couldn't have done it without them. But things are changing for us. We are so proud. You see, the farm has made it possible for us to send our two daughters to college! So now we have to think about maybe finding some new help for the farm as the girls go off on their own new lives."

"Wow, may I share your story with the Journal readers?" I asked.

"Oh. I don't know. I don't think so. It's really not that exciting a story. No, come to think of it. No, please don't tell anyone who we are or where we are from! Please? You see, we've been doing this for so long I almost forgot. We sell our raw milk to families who have been our friends and customers for years. We sell it with a label which says 'not for human consumption'. In this state it is strictly illegal to sell raw milk. We could get into trouble and so could our customers if the authorities found out," she said.

Then not so long ago I had another phone conversation with the wife of another farmer! Seven cows, fruits, vegetables, and honey. Modest sustained financial success accompanied with great satisfaction and general happiness.

Wow, three of them, spread out across the country! And then we got more stories from additional folks. Then I had to slap myself. Why was I surprised? Of all people. We have known for decades that the small family farm model is inherently powerful in its opportunity for success. We have, through the pages of the Small Farmer's Journal, championed just such adventures and

models. We know the numbers, we know how much income
might be realized from an acre of organically produced fruits and
vegetables (we're talking a wide range within five figures) while
sustaining or increasing fertility. We also know, from personal
experience, how much milk a cow will yield per day and every day
through lactation. And just how easy it is to sell that magnificent
true product direct to individuals (legally and otherwise). And
that it is possible with eight cows to realize an income, today, of
between $200 and $400 per day. We know these things. We
know, also, that the family milking those cows will enjoy on their
own table an assortment of the highest quality dairy products,
products which are in the purest sense UNAVAILABLE in any
supermarket in the western world. And that is not to mention the
other 'healthy' foods that that same farm provides for themselves.
And it is, of course, not limited to models centered around
milking cows. High quality meats, mushrooms, oil seed plants,
herbs, and so much more also offer many of the same dynamics.
We know these things. And we know there is no more satisfying
work available to humankind than that of the good farm. Working out of doors, with nature, fitting in with the biological
scheme of things, hands in the dirt, involved in generations of
plants and animals, the mind fully and harmoniously engaged in
observation - reaction - analysis - and creativity, peaceably growing food to feed people, all while protecting and revering our
little corner of the world. A work which defines a person in the
fullness of the best of humanity yet with all appropriate humility.
We know these things. So why is it that I feel, at this time, a
particular intangible peculiar illumination from the three models?
The answer is in those three little words "at this time."

at this time

Uncertainty permeates into most corners of our society.
Uncertainty about the future, about ethics, about our fellow
humans. To some, the artificially propped-up economy is a
serious worry, how long will it move in two competing directions? With increased abject poverty across the country and the

globe while a small percentage of folks enjoy an ever increasing
affluence, we are living in a time of dangerous social imbalance.
To others, without arguing the disparaties, we are doing great.
They see this, oddly, as evidence of our greatness. Around the
world and including in the U.S., 60,000 plus children die need-
lessly of hunger EVERY DAY and our own country regularly gets
in the way of her farmers providing food to those children because of the corporate perception that such a trend would reduce
profitability. Instead, our agribusiness community argues we are
the best at 'manufacturing' food because we employ the least
number of man hours per calorie produced! We feed the world by
example, they would seem to be arguing.

A huge majority of the world's scientific community is in
complete agreement on the subject of global warming. They say
the climate of the planet is changing rapidly and the storms we
have recently witnessed or suffered through will only increase in
number and intensity. They say that the world community can
do things to slow and even stop this climatological trend. Many
governments of the world are working in that direction. Our
government is not. What new climate related catastrophes await
us? Can our fragile, artificially propped-up economy withstand
much more of this?

Within the last six months I have read a dozen serious news
pieces from all over the U.S. and Europe presenting the possibility that we are entering the first stages of the end of oil. Whether
this happens or not, it has already become a self-fulfilling proph-
ecy as corporate greed and federal fear-mongering maneuver to
justify rapid increases in fuel prices. Anybody else notice that we
have dozens of people in the White House positioned to make
vast fortunes from the most recent fuel price increases? Mean-
while, what about the future of oil? Are intelligent questions
being asked about the short and long term consequences of an
end to oil?

Humankind has witnessed and experienced almost two years
of horrific natural disasters from floods, to earthquakes, to fires,
to famine, to pandemic disease, to pestilence, to poisonous foods,

to abject poverty, to hunger, to war and beyond. Yes, I include as *horrific natural disasters* poisonous foods, poverty, hunger and war. So much of what is happening in the world could have been avoided or softened had we been willing, before the fact, to trade greed and convenience for intelligent stewardship and planning.

Across all party lines and politcal persuasions, way out across the planet and back, our leaders - with few exceptions, are self-serving pragmatists, liars, cowards, idiots, and/or clever villains. And I am not just speaking of those at the top, the stupidity and cancerous greed permeates throughout. Because you see we, in most cases, put them there. We are part of the problem. You and I, if by no other factor than our misplaced tolerance and crime of silence. I don't know about you, but I've lost my tolerance and I don't feel like being quiet any longer.

I speak offensively and you should take offense. Leave if you must, but the time is past for sweet talk, for shoveling jargon, pretty rationales and the grease of compromise. There are those in business and politics who speak of the strength of the economy and the lovely inevitable future for freedom. And then there are those of us who think we are in for a world of hurt. I believe our economy is in trouble, the energy system is about to fall apart, our productivity is laughable, our relative levels of education are embarrassing, our health is under seige, we are at war at home and abroad, and we have no national governance and very little at the local level. And I care about all of that and more. Because I say these things I am branded a radical, a trouble maker, an agitator, and even accused of being unAmerican.

One evening at the 2005 U.S. plowing championship at Carriage Hill Farm, I sat under a tent with folk of Ohio and Maine discussing the future of farming. My friend, Arthur Bolduc of Ohio, sat next to me. At one point in the conversation a gentleman made the statement that union leadership would not allow a certain terrible thing to happen, at which Arthur responded;

"Our government, our unions and our churches have all failed us."

I could not agree more, though I realize to make such a statement in print, on these pages, invites bitter dispute. Even so, I would inflame the conversation even further and add that our schools and our political parties (all of them) must be added to the list.

There is no leadership at the top. None. The leadership is out arguably where it belongs, in our fields, in the barn milking those cows, tending the bees, preparing flocks for winter, in far flung and out of the way corners of man's best endeavors.

What are we to do? Dig in and get ready for one nasty roller coaster ride. And when I say dig in I mean deep and at the home place. Could I be wrong? You bet I could. But if a storm is coming right at us, I'd rather err on the side of caution rather than wanting to believe the storm will veer off at the last moment and leave me and my own unscathed.

Dig In? Yeah. It's time to return to basics. It's time to return to a rock hard valuation of self-sufficiency. Those who know how to raise and prepare their own foods, who know how to heat and house themselves, who understand and appreciate good tools and how to use them, they are the ones who will ride out most of the coming catastrophes. And they are the ones who should lead and by example.

what about you?

Hey, I know how to ask a horse to work for me in harness. I know how to harness him. I know how to hook him to farm tools and how to make those tools work right. That's pure gold today. I wouldn't trade what I know for any brand new tractor. Because I also know, from experience, that I can get my farming done pleasureably and profitably with horses in harness, and not be fretting about deisel or gasoline to get the crops in the ground and then harvested. Now that's the best kind of insurance. It's called <u>assurance</u>. Does it take me longer? Am I forced to farm fewer acres than my industrial counterpart? Where the heck are these questions coming from? If it takes me longer and I have fewer acres I'm most likely doing a better job of farming and keeping a whole lot more of the money at the end of the year. Do I really want to go faster on a bigger place and fall deeper in debt every year? The answer should be a resounding NO!

What about you? Can you tease and flirt with a piece of dirt in the right way so that crops result? If you had to put a piece of old steel in a coal fire and then bang it into a shape to repair a farm implement, could you do it? Would you be able to adapt your tractor to alternative fuel? If you came across a cow struggling to give birth would you know what to do? Would you know how to hook a team of mules to a mower and mow a crop of grass? Would you recognize problems in the look of a stream of milk as you milked your cow? Would you be able to eviscerate a bloody dead chicken?

What does any of this have to do with the condition of the world today? Everything. It is just such a litany of farming challenges, turned skills, which represents the best hope for society. You're right, not everybody will be able to do these things. But the world needs millions who can. And today's society creates high barriers to those who would go there by choice.

Our government thinks it needs a populace of followers, numb to life's best and most rewarding challenges, incapable of true self-sufficiency because that breeds free thinking. The independent small farmer is, almost by definition, solidly grounded by a marriage of instinct and creativity. She'll swim sideways for a while to test a companion planting or a breeding outcross or a harvest moment, but she always knows when its time to come up for air. And her moment for air will be different than the next person's moment for air. Doing things different, from other motivations, by other internal clocks, with other valuations, synchronized varigation, that's the bounty of independence. It makes for a constituency of one. Advertising, industry and government can't abide by constituencies of one. They depend on the mass of like-minded. The truly independent small farmer is seen as a thorn to modern America society when it should be seen for what it really is, a saviour of civilization. Because when you have a world made up of free thinking independent individuals you have the best and loveliest opportunity for local self-reliance and true collective self-sufficiency.

what's to be done?

My opinion; your choice whether or not to consider it:

If you are farming now you need to look at each and every way that you are dependent on the "grid": electricity, natural gas, fossil fuels, purchased inputs (seeds, supplies, industrial services, banking, etc.) and social ties (church, school, etc.).

• Ask yourself if you can get by when the electricity goes down for weeks on end, or when gasoline is rationed, or when your church or school shuts down.

• Ask yourself what happens when the local supermarket shelves aren't restocked.

• Ask yourself what happens if your bank should close its doors indefinitely, or the federal government order a banking holiday should too many people want their money at the same time.

• Ask yourself what happens when the local farm supply runs out of your seed, or fertilizer, or tractor parts.

• Ask yourself what it would mean if the government, for reasons of national security, should commandeer your production or your young adult offspring for some difficult to ratio-nalize federal war effort.

• Ask yourself what it would mean if trucks, trains, barges, or planes were not able to transport your production after harvest.

Ask yourself any such question which you feel contains a seed of insecurity for the future of your farm and family.

Follow the string on each of these considerations and figure out how you might prepare for such eventualities in ways that give you a solid sense of self-reliance. There are many ways to supplant or replace electricity. News is getting out of how thousands of folks are replacing refined diesel with biodiesel and even ordinary vegetable oils. Thousands of people are taking a deservedly hard look at the incredible practicality of true horsepower. The kitchen garden and greenhouse, along with reliable cold storage, will secure your family's food needs. And most of us are capable of projecting into the future just what supplies we might need for the farm. As for financial assets invested and on deposit, I'd be looking for some intelligent council from those who share a concern for instability.

When our farms are secured through a deliberate program of self-reliance we become a reservoir of strength and hope for those farm and non-farm neighbors who might not be so well prepared. When many of us do this together we build communities that will sustain and flourish. We know these things. Once, in the beginning of this nation, we took such considerations as formula for survival and growth.

If you want to be farming someday, start now, don't wait. The reason is that your farming dream may be the best way for you and your family to weather the coming storms. You may say, 'I haven't the money yet.' That doesn't have to stop you. There are many ways to get going with farming without a pile of money. First, understand that you don't have to own land to farm land. You can lease or rent ground where it is possible to grow crops, gardens, livestock. The possibilities are almost limitless. Take advantage of the great agricultural irony of our day, more and more exceptional farm land falls out of use each year, all around the country. While the real estate market booms with frightening, artificial, speculative valuations, non-farm folks flock to certain areas leaving others unwanted. If you insist on farming near Sedona, Arizona or Bend, Oregon, or on Martha's Vineyard, you're in for a real challenge. However, if you are willing to consider living and farming in the other 90% of the U.S., opportunities abound.

If you want to be farming someday, you also need to be giving careful consideration to how you depend on the extended 'grid' we spoke of earlier. And your questions will have a slightly different tilt. For example;

> • Ask yourself *where and how you might farm
> so that* you can get by when the electricity goes
> down for weeks on end, or when gasoline is
> rationed, or when your church or school shuts
> down.

You also have a broader consideration which should include what you will farm, what you will raise, how component aspects of a farming plan might be devised to add strength to certain inherent possibilities for self-reliance? I would like to recommend that, whether you are farming 10 acres or 300, you give careful thought to designing a mixed crop and livestock plan. Animal manures, properly gathered and applied, can reduce the need for purchased fertilizer inputs rather dramatically. Hogs on pasture can be creatively used to open new fields and beneficially restructure soils. Pastured poultry, while providing excellent protein sources for table and sale, can be called upon to reduce insect populations. Sheep, while providing wool and meat, are excellent for controlling grasses and weeds in small difficult to get to areas of the farm. Dairy and beef cattle will prefer pastures and crop residues that may differ from those enjoyed by horses, hogs and sheep. Many of the marginalized endangered livestock breeds are ideally suited for employ on small diversified farms. The use of horses or mules in harness as a power source for the farm is most definitely far-sighted and futuristic. For more reasons than we can list here, a judicious mix of livestock can be a golden center for the self-reliant farm of the twenty-first century.

Most important for you and for us is that you get on with your farming dream. The time is ripe right now. And, as uncertainty swells, the need for locally produced foods will only grow and grow. We need you farming as much you need to be farming.

specific intolerance

I wasn't around during the Depression. I was born right after World War II. My parents went through both of those times. As did many of my older friends. My father told me recently that these days, today - yesterday and tomorrow, worry him more than those two previous terrible times. He was a master sargent in the Marines in the thick of combat in the south Pacific during the Great War. Before that he was a teen-ager on a Wisconsin farm during the Great Depression. To paraphrase his recent advice to me he said, "You need to plan for the worst and take care of your own." I know he is right.

The worst might not happen, but I don't like the odds. If I go soft, and trust that things will work out, good chance I will be wrong and in worse shape for lack of preparation. Besides which, planning for self-sufficiency is a good thing. It fits in with most of the work I enjoy.

And those people we spoke of in the beginning of this writing? They are set! They are doing just fine and will continue. They are examples for each of us. Examples that show us, remind us, that we know these things. It's time to get back to basics. I, for one, get excited about our plans to get off the extended grid even further than we already are. It is invigorating to think of the farm or ranch as potentially self-reliant. It will be even more invigorating and assuring when it becomes an actuality.

The future is all yours, all ours. With each self-reliant farm we contribute to a better world. We work steadily, solidly, gradually towards less poverty, less hunger, less political insolvency, less manic consumption, and less environmental degradation. And, I might add, we work towards greater specific intolerance. I, for one, feel it is time to holler out 'no more business as usual!' There ain't no room in my world for the compromise which compromises, for the pragmatism born of the lie, for the political expediency which springs from conventional wisdom. There ain't no room for self-serving idiots at the top. There ain't no room for corporate ethics. There ain't no room for industrial foods. There ain't no room for cowardice. Now, every time I see a neck on an elected politician I see, and feel, the handle of a pitchfork. I feel my hands circle it as I pick it up with a load of beneficial manure and direct it towards the compost pile. That's what I mean by specific intolerance.

The future is ours but only if we take it by the handle. And we know these things...

Chapter Twenty-One

Sanctuary in the Measure

Early in the morning a view of the full barn reminds you of the bus load of junior high school football team members who came, with the coach, to help you, your husband and your two lovely daughters put up the hay before the summer storm. Coach said he'd never seen the boys work so hard. You'd never seen the girls work so hard, at the hay and at pretending to ignore the boys.

Today your husband left early to go to the neighborhood blacksmith shop to have the suction set on the plow point. He promised to stop by the Jensen's and pick up the six feeder pigs and those egg carton labels they printed for you, all in trade for that batch of setting hens. The girls are out bringing in the two teams of horses, the old reliable geldings and the young pair Tom's training for the Blount family. You fill the mangers and clean the barn in preparation. Horses stabled, you and the girls make the rounds; first the creep feeder for the one hundred lambs is filled - then their water is seen to. As you head to care for the hens and collect eggs, a car pulls up. It's the preacher and his wife on their once a week supply run. One daughter goes to tend to them. They've come to pick up their weekly 35 dozen of the brown eggs to take to the church bible school where moms will each collect their order. They also place an order for three 800 lb. crates of windfall apples to be delivered for the church cider pressing party. They take with them the new herb catalog sheets you've printed up. They leave behind money, smiles and a request that they be allowed to come with several families to help gather those windfall apples.

The egg count this morning is 267. The afternoon collection needs to be at least 100 to fill the orders. The new cafe in town has asked for an increase from 12 dozen to 20 dozen a day. They say the strong color, shape and flavor of the eggs has their customers bragging about the old days. They can get eggs from the restaurant supplier at 1/3 the price but they want yours, have even asked if they can add a little blurb on your farm to their breakfast menu.

You remember that morning a year ago when Tom said "we need 20 cents per egg to make this work" and you said "okay I'll get .20 per egg." $2400 dollars a month, half of which went into the girl's college fund. Even you were surprised how well it was working. Five hundred laying hens, five hundred poults getting ready to start laying and a batch of five hundred chicks coming up as replacement. The layers are sold just as soon as the poults begin to lay regular. Tom joked that the money from the sale of laying hens should be set aside for yours and his own education with field trips to exotic locales. But a line had been drawn in the sand: Tom said "The sheep, the orchard, the herbs and our plans for the work horses must come first. If this egg business takes off, we need to agree that it will get no bigger than what we have here. We don't want it taking over and making us into some industrial egg plant." You knew he was right and agreed before his speech was made.

It was the herb patch which was creating the most excitement this season. You both weren't sure when you switched to personal guarantees and warranties on your organic labeling how it would affect sales. The letter from the USDA had been particularly nasty, but the attorney, paid with cut and wrapped warranteed organic lamb, had quieted the bureaucrats for now and the article in the newspaper had brought a whole bunch of new customers. There would be some challenges soon on how to dry and bag larger quantities, but Tom had a great idea for a thirty foot long revolving wire 'sandwich' tray.

Your biggest challenge will come when the girls go off to start their own lives. You have talked long and hard about how to 'maybe' get one or both of them to want to stay on, or return to the farm. The outside world has its attractions, but you are certain that the 'ownership' you've given the girls of many aspects of the place and the work should give the best chance of their return with, it is hoped, young families of their own. They have each picked out spots on the farm where they'd like to see their own future homes built.

In the beginning you thought it was unnecessary to include the girls in the twice a week evening budget talks with Tom. But you have to admit he was right, it has been very good for the young women to learn why and how certain choices are made. And their input has been useful and exciting. It was Jenny who suggested that time could be saved daily by building the short fence which created the lane giving the work horses a way to come, on their own, closer to the barn. Now instead of walking a quarter mile to halter and lead the horses back, usually they were waiting within 200 feet of the barn for halters and an open gate. She was proud and seemed to take credit for how the old team went straight to their own stall without being led. She timed herself and claimed 20 minutes of her time saved each work day, almost 6 hours a month! The math had spurred Tom to all sorts of calculations of how to be more efficient. He liked to say that Jenny had shown how they could gain a free farm worker just by saving steps.

Overall things on the farm are going quite well. Income is growing, especially when all the true value of the bartering is factored in. It had been your idea that each thing traded off or traded for should have an assigned value and be entered into the book work. Tom said, initially, 'No, that wasn't real money'. And you pointed out that it would have cost real money to pay to print the egg carton labels and those setting hens were worth real money. By factoring in all the less tangibles (including the girls' labor paid as it was at below the market value, including the value of home consumed food, including dramatically reduced doctor's bills, including all the labor and service trades) your 120 acre farm had brought in just under 200,000 dollars gross last year. Had only the cash in hand been counted, the gross revenue would have registered as 110,000 dollars. You know these numbers mean less to you than the experience of living on, working with and sharing this right livelihood with your family. For that there is no dollar measurement.

Chapter Twenty-Two

The Farmer's Cafe - Terra Ortus

"If you will cling to Nature, to the simple in Nature, to the little things that hardly anyone sees, and that can so unexpectedly become big and beyond measuring; if you have this love of inconsiderable things and seek quite simply, as one who serves, to win the confidence of what seems poor: then everything will become easier, more coherent and somehow more conciliatory for you, not in your intellect, perhaps, which lags marveling behind, but in your inmost consciousness, waking and cognizance."
                                                              Rainer Maria Rilke

"Love is, among other things, the experience of wholly identifying with another person's sincerety."
                                                              Peter Schjeldahl

For the mass of us, the connection between food and farming is dry, distant, awkward, academic. We do not associate the squeaking sticky wet-sand dig with the flavor, smell and texture of clams because we buy them off the shelf. We do not associate the crystalline departing cold and invasive spring warmth nor the apiarian ballet amongst bouquets of apple blossoms with the resultant fruit in the supermarket bins. Most of us cannot, no CAN NOT, no WILL NOT, be bothered by the terrible tactile truths of where and how our meats come to us. And many of us haven't a clue. We are separated, artificially, by contrived modern circumstance, from all that would define us. Ah, but if we only knew what meaning we deny ourselves, what music, what dance steps, what loveliness, what purpose. Eyes, mind, pores and hands open we may actually borrow from the waiting backside of tomorrow a vital warmth.

There is the potential, realized in many small cultures, that every meal spring from the living life of all its components, most certainly and definitively including the food preparer and the consumer. A handful of people know of a rare and superior variety of melting cheese which is produced from the milk of an obscure breed of cattle. And only from the milk of French August. And only from those cows which pastured a certain high elevation legume mix on certain mineral rich soil. The songs may even go so far as to suggest that the cattle were called by a brass and glass throaty singing female call and that those same young women were humming while they milked. The resulting cheese, they would tell you, can never be fully understood in any scientific way. But the people of this cheese see and taste in it not only the mountain pastures and the cattle but the pastures of the past and those possible in the future. They see in the cheese what must have been the look of their ancestral grandmother as she hummed to the lactating ancestral grandmother of today's cow. And they see and feel and taste the specific historical miseries, terrors and losses of that same living landscape in that cheese. There should be a word for this, we need a word available to us which speaks to all these connections, all this history, all this value. The fact that we do not have, in common use, this word, tells us we have lost this connection.

The battlefield of our near future will be the countertops and academic journals of the life sciences. It will become, has in some respects already become, the moral and ethical equivalent of a dirty unwelcome war. Humanity does not belong in this war with nature, she cannot win control over genetic codes and the deeper mysteries of life. Many of us do not want a part of this war. As with that French cheese, the power and value of what we identify as a living piece of us, our place in a culture of absolute interdependence with all of biology, is tied to the implicit fact that we must honor what we don't know while working to protect it. The war is set to make of us social heretics if not actual outlaws. Heretics because we dare not believe that science in service to business will deliver us from the evils of an untamed biological universe. Outlaws because we insist on keeping heirloom seeds and livestock breeds when told this is illegal. In spite of the posture we find ourselves in, one of fear and loathing, it is difficult but necessary to avoid being drawn into the battle.

We must understand that the only way we will keep a "living life" alive is to keep it to ourselves, actually and metaphorically. All of the vital connective tissues of our recipes and formulas and remembered procedures must be guarded and held in buried jars and secret coded texts. We can never succeed politically, legally or scientifically to end the war against the sanctity of life because we fight the larger arrogance of the dehumanized powerful few who believe they will improve biology. The only way we will win this war is to make our secret knowledge something mankind comes to realize it cannot live without. Until then we must keep the names we call our sheep, seeds, farming methods, compost recipes and each other secret. Otherwise these things may be stripped from us. We need to hide our true farming and smile while we accept the heapings of ridicule. We must speak only in fanciful fictionesque story-like recipes laced with unforgivable romance and ....

Poached Chicken in a Pot - terra ortus
- with apologies to LaVarenne Pratique.

This is a complete meal, the cooking broth may be served as soup, followed by the stuffed chicken with vegetables. Best served where all who partake may have a view of gardens, trees and fields. This recipe serves eight adult appetites magnificently IF they have worked in that garden, field and kitchen. However if those served are able-bodied yet indolent pleasure seekers, the layers, transparencies and textures of the meal's ingredients, (along with their histories) will be diluted and somewhat wasted.

6 - 9 pound stewing chicken
(preferrably the Patridge Cochin or Black Astralorp hen which never laid an egg and was always first at the feed trough. You or your mate have butchered and cleaned this bird.)

1 onion studded with two cloves. (The *Borrettana Cipollini* onion came from your kitchen garden. The cloves from a jar in the pantry where you keep superb goodies you have traded for from pen pal farmers in other climates.)

Large *bouquet garni*: preferably fresh (but no older than one year, carefully dried to retain color and aroma) your daughter's herb garden production of thyme, parsley, a bay leaf and a piece of leek green.

1 teaspoon black peppercorns
1 stalk of garden-grown Afina celery stalk, cut in pieces by the middle son.

1 cinnamon stick from that pantry of bartered treasures.
1 teaspoon of salt.
5 quarts (or more) of water

For the Stuffing
4 slices of dry white homemade oven-crusted bread (which the children kneaded and Daddy added Rosemary to).

1 cup of sheep, goat or Ayshire cow's milk preferably fresh squeezed and directly after the animal had pastured the easterly slope of a field rich in legumes growing tight in mineral rich soil where your grandparents are buried.

1 teaspoon of butter from a batch churned by a young woman in the blush of new love and humming with every stroke.

1/2 of a Bennie's Red onion from the garden and chopped.
The heart and liver of the departed hen.

1/2 pound of raw or cooked smoked ham, ground. This from the supply Uncle Joe provided from his own picturesque Hampshire hog farm just down the road.

1 clove of Spanish garlic, chopped and sung over with a Cuban lullaby.

4 tablespoons of chopped Gigante d'Italia parsely from the plant in the kitchen window box.

1 pinch of grated nutmeg from the jewels in the pantry.

salt and pepper

1 hand-gathered egg from the chicken house, beaten to mix.

Vegetable Garnish

1 1/2 pounds of medium Thumbelina carrots, trimmed. These are the ones which didn't "look" quite right to make it to the Saturday farmer's market.

2 pounds of garden fresh Siegfried leeks, trimmed and split.

1 pound of small White Egg turnips from that small field planted to flush out the pregnant ewes.

For serving

1/4 pound of very fine handmade noodles using cousin Larry's wheat.

To make the stuffing: soak mother's wonderful dry bread in the magic milk for 10 minutes, then squeeze it dry and crumble it. Melt the romantic butter and fry the fragrant onion until soft. Add the stolen chicken liver and heart and saute for 1 to 2 minutes until brown but still pink in the center. Cool slightly while humming, then chop the mixture and guide your children in stirring this into the breadcrumbs with the ham, garlic, parsley, nutmeg, salt and pepper to taste. Stir in the egg to bind the stuffing.

Stuff the bird's cavities and tie them closed. Tie the onion, bouquet garni, peppercorns, celery, and cinnamon in cheesecloth. Think about those loved ones you will miss at today's table and rehearse a poetic toast to them. Put the chicken in a pot with the cheesecloth bag, salt and water to cover, and bring slowly to a boil, skimming often. Send the children and guests to some far corner to do chores or games while you hold your loved one within the kitchen smells and whisper. Simmer uncovered for 1 hour, skimming occasionally. Taste frequently for your own good and with sipping wine or cream sherry. Pull out the plan book and work on ideas for next year's kitchen garden making sure to add a new variety of garlic and a hinged trellis for the peas.

For the garnish: tie the carrots and leeks in bundles and add them with the turnips. Simmer for another hour or until both chicken and vegetables are very tender. Add more water if necessary so that the bird is always covered. While tending the pot, sketch designs for labels to go on preserve jars destined for the farmer's market.

Take the chicken out of the stew pot and put it on the cutting board your father made for you. Throw away the trussing strings. Reduce the broth until it is well flavored. Carve the chicken and arrange the pieces artfully on the deep platter your daughter made and decorated. Pile the stuffing on top and the vegetables around the chicken, then cover the dish with foil and keep warm until ready to serve.

A short time before serving, spoon about 1 quart broth into a pan and simmer the noodles for 5 minutes or until tender. Throw away the cheesecloth bag and skim off as much fat as possible. Serve the tureen of noodle soup with the chicken and vegetables. At the dinner table take time to breathe and look, inconspicuously, at each person seated there while thinking of the soil from which all of this has sprung.

Sustenance survival fuel energy
taste communion fellowship occasion

Though we must, as guerilla farmers, protect through concealment what we are about, for the time being there seems little harm in, and some value from, identifying and discussing the dragon's breath.

We have heard and permitted ourselves to say time and again that sad cliche, "It's a small world." It is as if we are either sad to discover this, or that there are fewer surprises, or that *this* **is** all there is, or that we are worried to discover ourselves to seem bigger than what the world offers us, or that we should align ourselves to accept the small world ....

It is not a small world. Our synthetic western societal coccoon is made to give us the illusion of *limited*. It is far from true. Rather, it is a wide vast unfathomable world. With reasons, and acids, and flavors and outcomes beyond any single or collective imagination. We accept our world as small, tasteless, colorless, anitseptic, grey, predictable, unfortunately safe, cornerless, dry, drivable, fat free, friendly, immediate, and convenient, because we are doped dopes. Our electronic cyberspace needs us to believe it contains ALL the information we would ever need, our electronically-dependent supermarkets need us to believe they contain all the food we would want to eat. Our clever ad agencies need us to believe that their mind-altering tactics are in our best consuming interests. Our videographers depend on the capsule view. Our corporate government needs us to believe that it holds principled concern for our well being. Our corporate boardrooms need us to believe that nothing is more important nor beneficial to all than profitability.

Not only is the world not small, it has a nearly infinite capacity for biological, cultural, and spiritual growth. Whether *world* means for us our biological universe, our cultural tapestry, or the distances between recognized points, outside of the synthetic western coccoon of limited-life menus, every adventure, every life breeds new biology, culture and distance.

It is a supreme irony which finds corporate commercial *dictum* molding the gluttony of our modern affluence towards a lifeless pedantic moronic tasteless homogenized consumption of inane shallow little sensations.

At a time when many people should be able to marry education, sophistication and the means to appreciate and SHARE an unqualified passion for the whole of life, that whole which under the lens would give us ownership of a consumate passion for our food - a passion which demands an ever vigilant protection of the sanctity and vitality and diversity of that same food - instead we come as confinement hogs to the offal and chemical-laced troughs of the corporate board rooms.

There are societies thriving today where the history of

their crops and livestock, the history of their fields, the history of their farmers, the history of their harvests and storage, the history of their food preparation, and the history of their very meals are at the core of who each individual in that society is. All that is their farming from seed to supper is of their individual and collective essence, it is their soul.

Huevos Reales - Royal Eggs
with apologies to Mexico the Beautiful

As the story goes; the Dominican nuns of Mexico's Santa Rosa convent used egg whites to paint the convent walls. This dessert recipe was invented to make good use of the remaining yolks.

10 egg yolks
2 teaspoons baking powder
1 teaspoon butter
1/3 cup muscat raisins
cinnamon shavings for garnish

Oven at 275⁰. In large bowl beat yolks til thick and creamy. Stir in baking powder.

Grease 13 x 9 in. baking dish with butter & pour in egg mix. Cover with foil and bake for 45 minutes or until toothpick comes clean. Remove and cool for 10 min. To make syrup, mix the sugar, water and cinnamon in a small heavy saucepan and boil, stirring, for 5 min. until the mix forms a light syrup. Remove from heat and add sherry.

Cut eggs in dish to 1 in. squares. Cover with syrup and garnish with raisins and cinnamon shavings. Serve chilled or at room temperature.

For some of us the connection between food and farming is moist, near, comfortable, and holy. We associate flavors, smells, textures and process, back and forth, between the fields and the kitchen, the table and the barn, the future and the past, the family and the community. It is through this example, even with our secrets intact, that we have profound effect on everyone around us. No one is completely immune from the romance of nuns valuing egg whites to paint the convent walls and appreciating the remaining yolks for the sweets they bring. Wave it away as so much silliness, but the imprint on the soul made it through your defenses. It is in these ways we are true guerrilla farmers, in these ways we are the pirate chefs of the farmer's cafe. Eat smile and wink. The strength is yours. LRM

Afterword

The winnowed bright of dusty fiddling sets me against my tedious hope that for a correct static moment the picture would match the contrived anecdote and I would find myself looking back at not a perfect farm nor a perfect painting studio nor a perfect writing table but the beginning of a perfect day, something which must always belong to the sizing-up of a past moment.

A precarious balance is the circumstance that would allow a keen individual moments of a working appreciation for his or her transitory aspect. Writing of, for, and about my chosen work and workings sometimes gives me that keenness I crave. Perhaps I am to find that the efforts had the tooth and texture adequate to "size-up" for me, in a reflective moment, as the beginning of a perfect day. Maybe I will remember it all as part of a nearly perfect life.

An undeniable arrogance precedes my many attempts, over thirty some odd years, to put to the written word a set of observations and concerns, out away from me, where they would walk off and multiply and/or corrode.

If it is true that we become our work, and that the convictions are want to morph into difficult cloth, I am a circuit-riding preacher brandishing inapproriate weapons belted and holstered over a long dark coat which conceals an island print shirt. But what I am or what I am not sloughs away if the word pictures and the arguments have any power to permeate, to worm, to soothe, to tickle or to win over. I don't matter, the words do, the paintings do, the farm does. It is my suspicions which spoil the picture. Suspicions which cause me to doubt that the words, paintings and farm also do indeed matter. That perhaps the only thing which matters is some small evidence of the nature of the energy which was spent. Is this why we, I, condescend and talk down to our very selves?

Struggling to let go of it...

Lynn Miller
Singing Horse Ranch
Geneva, Oregon 2006

about the author

Lynn Ralph Miller lives on and operates a remote Oregon ranch with his wife and daughter. In 1976 he began the *Small Farmer's Journal,* an award-winning international agrarian quarterly. He continues as its editor/publisher. He is the author of over a dozen books, many on the subject of animal-powered agricultural systems. Mr. Miller has been actively involved with small farm politics and organic systems since 1968. For over thirty years he has chosen to depend on draft horses in harness as motive power for his farming and ranching, and is generally recognized as a world authority on the subject. On the subject of small farms and animal power, Mr. Miller has lectured and conducted workshops for decades all across the U.S. and Canada. He also holds degrees in the fine arts and for over forty years has drawn and painted. His art works are in scattered private collections throughout the U.S. and Europe.